John Tyndall

Fragaments of Science: a Series of Detached Essays, Addresses, and

Reviews

Vol. 1

John Tyndall

Fragaments of Science: a Series of Detached Essays, Addresses, and Reviews
Vol. 1

ISBN/EAN: 9783744718363

Printed in Europe, USA, Canada, Australia, Japan

Cover: Foto ©Thomas Meinert / pixelio.de

More available books at **www.hansebooks.com**

FRAGMENTS OF SCIENCE

A SERIES OF DETACHED
ESSAYS, ADDRESSES, AND REVIEWS

BY

JOHN TYNDALL, F. R. S.

AUTHOR OF NEW FRAGMENTS, HEAT AS A MODE OF MOTION,
ON SOUND, RADIANT HEAT, ON FORMS OF WATER,
HOURS OF EXERCISE IN THE ALPS, ETC.

VOL. I

NEW YORK
D. APPLETON AND COMPANY
1897

PUBLISHERS' NOTE.

THE first edition of Prof. Tyndall's "Fragments of Science" was published some twenty years ago as a single volume, which was made up of a score or more of his detached essays, addresses, and reviews. The book was afterward revised, some of the papers recast, and from time to time new ones added, until, the size of the work becoming somewhat unwieldy, the present two-volume edition was decided upon. This contains fifteen additional papers, and represents the author's latest changes and revisions. The volumes are uniform with "New Fragments," recently issued, and the three together include all the occasional writings which their author has decided to preserve in permanent form.

PREFACE

THE SIXTH EDITION.

To AVOID unwieldiness of bulk this edition of the ' Fragments ' is published in two volumes, instead of, as heretofore, in one.

The first volume deals almost exclusively with the laws and phenomena of matter. The second trenches upon questions in which the phenomena of matter interlace more or less with those of mind.

New Essays have been added, while old ones have been revised, and in part recast. To be clear, without being superficial, has been my aim throughout.

In neither volume have I aspired to sit in the seat of the scornful, but rather to treat the questions touched upon with a tolerance, if not a reverence, befitting their difficulty and weight.

Holding, as I do, the nebular hypothesis, I am logically bound to deduce the life of the world from forces inherent in the nebula. With this view, which is set forth in the second volume, it seemed but fair to associate the reasons which cause me to conclude that every attempt made in our day to generate life independently of antecedent life has utterly broken down.

A discourse on the Electric Light winds up the second volume. The incongruity of its position is to be referred to the lateness of its delivery.

CONTENTS

OF

THE FIRST VOLUME

———◆◆———

X CONTENTS.

MAP

INORGANIC NATURE

I.

THE CONSTITUTION OF NATURE.[1]

WE cannot think of space as finite, for wherever in imagination we erect a boundary, we are compelled to think of space as existing beyond it. Thus by the incessant dissolution of limits we arrive at a more or less adequate idea of the infinity of space. But, though compelled to think of space as unbounded, there is no mental necessity compelling us to think of it either as filled or empty; whether it is so or not must be decided by experiment and observation. That it is not entirely void, the starry heavens declare; but the question still remains, Are the stars themselves hung in vacuo? Are the vast regions which surround them, and across which their light is propagated, absolutely empty? A century ago the answer to this question, founded on the Newtonian theory, would have been, 'No, for particles of light are incessantly shot through space.' The reply of modern science is also negative, but on different grounds. It has the best possible reasons for rejecting the idea of luminiferous particles; but, in support of the conclusion that the celestial spaces are occupied by matter, it is able to offer proofs

[1] 'Fortnightly Review,' 1865, vol. iii. p. 129.

almost as cogent as those which can be adduced of the existence of an atmosphere round the earth. Men's minds, indeed, rose to a conception of the celestial and universal atmosphere through the study of the terrestrial and local one. From the phenomena of sound, as displayed in the air, they ascended to the phenomena of light, as displayed in the *ether*; which is the name given to the interstellar medium.

The notion of this medium must not be considered as a vague or fanciful conception on the part of scientific men. Of its reality most of them are as convinced as they are of the existence of the sun and moon. The luminiferous ether has definite mechanical properties. It is almost infinitely more attenuated than any known gas, but its properties are those of a solid rather than of a gas. It resembles jelly rather than air. This was not the first conception of the ether, but it is that forced upon us by a more complete knowledge of its phenomena. A body thus constituted may have its boundaries; but, although the ether may not be co-extensive with space, it must at all events extend as far as the most distant visible stars. In fact it is the vehicle of their light, and without it they could not be seen. This all-pervading substance takes up their molecular tremors, and conveys them with inconceivable rapidity to our organs of vision. It is the transported shiver of bodies countless millions of miles distant, which translates itself in human consciousness into the splendour of the firmament at night.

If the ether have a boundary, masses of ponderable matter might be conceived to exist beyond it, but they could emit no light. Beyond the ether dark suns might burn; there, under proper conditions, combustion might be carried on; fuel might consume unseen, and metals be fused in invisible fires. A body, moreover, once heated there, would continue for ever heated; a sun or planet once molten, would continue for ever molten. For, the loss of heat being simply the abstraction of molecular motion by the ether, where this medium is absent no cooling could occur. A sentient being on approaching a heated body in this region, would be conscious of no augmentation of temperature. The gradations of warmth dependent on the laws of radiation would not exist, and actual contact would first reveal the heat of an extra ethereal sun.

Imagine a paddle-wheel placed in water and caused to rotate. From it, as a centre, waves would issue in all directions, and a wader as he approached the place of disturbance would be met by stronger and stronger waves. This gradual augmentation of the impression made upon the wader is exactly analogous to the augmentation of light when we approach a luminous source. In the one case, however, the coarse common nerves of the body suffice; for the other we must have the finer optic nerve. But suppose the water withdrawn; the action at a distance would then cease, and, as far as the sense of touch is concerned, the wader would be first rendered conscious of the motion of the wheel by the blow of the paddles. The transference of motion from the paddles to the water is mechanically similar to the transference of molecular motion from the heated body to the ether; and the propagation of waves through the liquid is mechanically similar to the propagation of light and radiant heat.

As far as our knowledge of space extends, we are to conceive it as the holder of the luminiferous ether, through which are interspersed, at enormous distances apart, the ponderous nuclei of the stars. Associated with the star that most concerns us we have a group of dark planetary masses revolving at various distances round it, each again rotating on its own axis; and, finally, associated with some of these planets we have dark bodies of minor note—the moons. Whether the other fixed stars have similar planetary companions or not is to us a matter of pure conjecture, which may or may not enter into our conception of the universe. But probably every thoughtful person believes, with regard to those distant suns, that there is, in space, something besides our system on which they shine.

From this general view of the present condition of space, and of the bodies contained in it, we pass to the enquiry whether things were so created at the beginning. Was space furnished at once, by the fiat of Omnipotence, with these burning orbs? In presence of the revelations of science this view is fading more and more. Behind the orbs, we now discern the nebulæ from which they have been condensed. And without going so far back as the nebulæ, the man of science can prove that out of common non-luminous matter this whole pomp of stars might have been evolved.

The law of gravitation enunciated by Newton is, that every particle of matter in the universe attracts every other particle with a force which diminishes as the square of the distance increases. Thus the sun and the earth mutually pull each other; thus the earth and the moon are kept in company; the force which holds every respective pair of masses together being the integrated force of their component parts. Under the operation of this force a stone falls to the ground and is warmed by the shock; under its operation meteors plunge into our atmosphere and rise to incandescence. Showers of such meteors doubtless fall incessantly upon the sun. Acted on by this force, the earth, were it stopped in its orbit to-morrow, would rush towards, and finally combine with, the sun. Heat would also be

developed by this collision. Mayer first, and Helmholtz and Thomson afterwards, have calculated its
amount. It would equal that produced by the combustion of more than 5,000 worlds of solid coal, all
this heat being generated at the instant of collision.
In the attraction of gravity, therefore, acting upon
non-luminous matter, we have a source of heat more
powerful than could be derived from any terrestrial
combustion. And were the matter of the universe
thrown in cold detached fragments into space, and there
abandoned to the mutual gravitation of its own parts,
the collision of the fragments would in the end produce the fires of the stars.

The action of gravity upon matter originally cold
may, in fact, be the origin of all light and heat, and
also the proximate source of such other powers as are
generated by light and heat. But we have now to
enquire what is the light and what is the heat thus
produced? This question has already been answered
in a general way. Both light and heat are modes of
motion. Two planets clash and come to rest; their
motion, considered as that of masses, is destroyed, but
it is in great part continued as a motion of their
ultimate particles. It is this latter motion, taken up
by the ether, and propagated through it with a velocity of 186,000 miles a second, that comes to us as
the light and heat of suns and stars. The atoms of a
hot body swing with inconceivable rapidity—billions of
times in a second—but this power of vibration necessarily implies the operation of forces between the
atoms themselves. It reveals to us that while they
are held together by one force, they are kept asunder
by another, their position at any moment depending
on the equilibrium of attraction and repulsion. The
atoms behave as if connected by elastic springs, which

2

oppose at the same time their approach and their retreat, but which tolerate the vibration called heat. The molecular vibration once set up is instantly shared with the ether, and diffused by it throughout space.

We on the earth's surface live night and day in the midst of ethereal commotion. The medium is never still. The cloud canopy above us may be thick enough to shut out the light of the stars; but this canopy is itself a warm body, which radiates its thermal motion through the ether. The earth also is warm, and sends its heat-pulses incessantly forth. It is the waste of its molecular motion in space that chills the earth upon a clear night; it is the return of thermal motion from the clouds which prevents the earth's temperature, on a cloudy night, from falling so low. To the conception of space being filled, we must therefore add the conception of its being in a state of incessant tremor.

The sources of this vibration are the ponderable masses of the universe. Let us take a sample of these and examine it in detail. When we look to our planet, we find it to be an aggregate of solids, liquids, and gases. Subjected to a sufficiently low temperature, the two last would also assume the solid form. When we look at any one of these, we generally find it composed of still more elementary parts. We learn, for example, that the water of our rivers is formed by the union, in definite proportions, of two gases, oxygen and hydrogen. We know how to bring these constituents together, so as to form water : we also know how to analyse the water, and recover from it its two constituents. So, likewise, as regards the solid portions of the earth. Our chalk hills, for example, are formed by a combination of carbon, oxygen, and calcium. These are the so-called *elements* the union of which, in definite proportions, has resulted in the formation of chalk. The flints within the chalk we know to be a compound of oxygen and silicium, called silica; and our ordinary clay is, for the most part, formed by the union of silicium, oxygen, and the well-known light metal, aluminium. By far the greater portion of the earth's crust is compounded of the elementary substances mentioned in these few lines.

The principle of gravitation has been already described as an attraction which every particle of matter, however small, exerts on every other particle. With gravity there is no selection ; no particular atoms choose, by preference, other particular atoms as objects of attraction ; the attraction of gravitation is proportional simply to the quantity of the attracting matter, regardless of its quality. But in the molecular world which we have now entered matters are otherwise arranged. Here we have atoms between which a strong attraction is exercised, and also atoms between which a weak attraction is exercised. One atom can jostle another out of its place, in virtue of a superior force of attraction. But, though the amount of force exerted varies thus from atom to atom, it is still an attraction of the same mechanical quality, if I may use the term, as that of gravity itself. Its intensity might be measured in the same way, namely by the amount of motion which it can generate in a certain time. Thus the attraction of gravity at the earth's surface is expressed by the number 32 ; because, when acting freely on a body for a second of time, gravity imparts to the body a velocity of thirty-two feet a second. In like manner the mutual attraction of oxygen and hydrogen might be measured by the velocity imparted to the atoms in their rushing together. Of course such a unit of time as a second is not here to be thought of, the whole interval required by the atoms to cross the minute spaces which separate them amounting only to an inconceivably small fraction of a second.

It has been stated that when a body falls to the earth it is warmed by the shock. Here, to use the terminology of Mayer, we have a *mechanical* combination of the earth and the body. Let us suffer the falling body and the earth to dwindle in imagination to the size of atoms, and for the attraction of gravity let us substitute that of chemical affinity; we have then what is called a *chemical* combination. The effect of the union in this case also is the development of heat, and from the amount of heat generated we can infer the intensity of the atomic pull. Measured by ordinary mechanical standards, this is enormous. Mix eight pounds of oxygen with one of hydrogen, and pass a spark through the mixture; the gases instantly combine, their atoms rushing over the little distances which separate them. Take a weight of 47,000 pounds to an elevation of 1,000 feet above the earth's surface, and let it fall; the energy with which it will strike the earth will not exceed that of the eight pounds of oxygen atoms, as they dash against one pound of hydrogen atoms to form water.

It is sometimes stated that gravity is distinguished from all other forces by the fact of its resisting conversion into other forms of force. Chemical affinity, it is said, can be converted into heat and light, and these again into magnetism and electricity: but gravity refuses to be so converted; being a force maintaining itself under all circumstances, and not capable of disappearing to give place to another. The statement arises from vagueness of thought. If by it be meant that a particle of matter can never be deprived of its weight, the assertion is correct; but the law which affirms the convertibility of natural forces was never intended, in the minds of those who understood it, to affirm that such a conversion as that here implied occurs in any case whatever. As regards convertibility into heat, gravity and chemical affinity stand on precisely the same footing. The attraction in the one case is as indestructible as in the other. Nobody affirms that when a stone rests upon the surface of the earth, the mutual attraction of the earth and stone is abolished; nobody means to affirm that the mutual attraction of oxygen for hydrogen ceases, after the atoms have combined to form water. What is meant, in the case of chemical affinity, is, that the pull of that affinity, acting through a certain space, imparts a motion of translation of the one atom towards the other. This motion is *not* heat, nor is the force that produces it heat. But when the atoms strike and recoil, the motion of translation is converted into a motion of vibration, which *is* heat. The vibration, however, so far from causing the extinction of the original attraction, is in part carried on by that attraction. The atoms recoil, in virtue of the elastic force which opposes actual contact, and in the recoil they are driven too far back. The original-attraction then triumphs over the force of recoil, and urges the atoms once more together. Thus, like a pendulum, they oscillate, until their motion is imparted to the surrounding æther; or, in other words, until their heat becomes *radiant* heat.

In this sense, and in this sense only, is chemical affinity converted into heat. There is, first of all, the attraction between the atoms; there is, secondly, *space* between them. Across this space the attraction urges them. They collide, they recoil, they oscillate. There is here a change in the form of the motion, but there is no real loss. It is so with the attraction of gravity. To produce motion by gravity space must also intervene between the attracting bodies. When they strike to-

gether motion is apparently destroyed, but in reality
there is no destruction. Their atoms are suddenly urged
together by the shock; by their own perfect elasticity
these atoms recoil; and thus is set up the molecular
oscillation which, when communicated to the proper
nerves, announces itself as heat.

It was formerly universally supposed that by the collision of unelastic bodies force was destroyed. Men saw,
for example, that when two spheres of clay, painter's
putty, or lead for example, were urged together, the
motion possessed by the masses, prior to impact, was
more or less annihilated. They believed in an absolute
destruction of the force of impact. Until recent times,
indeed, no difficulty was experienced in believing this,
whereas, at present, the ideas of force and its destruction refuse to be united in most philosophic minds. In
the collision of elastic bodies, on the contrary, it was
observed that the motion with which they clashed together was in great part restored by the resiliency of
the masses, the more perfect the elasticity the more
complete being the restitution. This led to the idea of
perfectly elastic bodies—bodies competent to restore by
their recoil the whole of the motion which they possessed
before impact—and this again to the idea of the *conservation* of force, as opposed to that destruction of
force which was supposed to occur when unelastic bodies
met in collision.

We now know that the principle of conservation
holds equally good with elastic and unelastic bodies. Perfectly elastic bodies would develop no heat on collision.
They would retain their motion afterwards, though its
direction might be changed; and it is only when sensible
motion is wholly or partly destroyed, that heat is generated. This always occurs in unelastic collision, the
heat developed being the exact equivalent of the sensible

motion extinguished. This heat virtually declares that the property of elasticity, denied to the masses, exists among their atoms; by the recoil and oscillation of which the principle of conservation is vindicated.

But ambiguity in the use of the term 'force' makes itself more and more felt as we proceed. We have called the attraction of gravity a force, without any reference to motion. A body resting on a shelf is as much pulled by gravity as when, after having been pushed off the shelf, it falls towards the earth. We applied the term force also to that molecular attraction which we called chemical affinity. When, however, we spoke of the conservation of force, in the case of elastic collision, we meant neither a pull nor a push, which, as just indicated, might be exerted upon inert matter, but we meant force invested in motion—the *vis viva*, as it is called, of the colliding masses.

Force in this form has a definite mechanical measure, in the amount of work that it can perform. The simplest form of work is the raising of a weight. A man walking up-hill, or up-stairs, with a pound weight in his hand, to an elevation say of sixteen feet, performs a certain amount of work, over and above the lifting of his own body. If he carries the pound to a height of thirty-two feet, he does twice the work; if to a height of forty-eight feet, he does three times the work; if to sixty-four feet, he does four times the work, and so on. If, moreover, he carries up two pounds instead of one, other things being equal, he does twice the work; if three, four, or five pounds, he does three, four, or five times the work. In fact, it is plain that the work performed depends on two factors, the weight raised and the height to which it is raised. It is expressed by the product of these two factors.

But a body may be caused to reach a certain elevation in opposition to the force of gravity, without being actually carried up. If a hodman, for example, wished to land a brick at an elevation of sixteen feet above the place where he stood, he would probably pitch it up to the bricklayer. He would thus impart, by a sudden effort, a velocity to the brick sufficient to raise it to the required height; the work accomplished by that effort being precisely the same as if he had slowly carried up the brick. The initial velocity to be imparted, in this case, is well known. To reach a height of sixteen feet, the brick must quit the man's hand with a velocity of thirty-two feet a second. It is needless to say, that a body starting with any velocity, would, if wholly unopposed or unaided, continue to move for ever with the same velocity. But when, as in the case before us, the body is thrown upwards, it moves in opposition to gravity, which incessantly retards its motion, and finally brings it to rest at an elevation of sixteen feet. If not here caught by the bricklayer, it would return to the hodman with an accelerated motion, and reach his hand with the precise velocity it possessed on quitting it.

An important relation between velocity and work is here to be pointed out. Supposing the hodman competent to impart to the brick, at starting, a velocity of sixty-four feet a second, or twice its former velocity, would the amount of work performed be twice what it was in the first instance? No; it would be four times that quantity; for a body starting with twice the velocity of another, will rise to four times the height. In like manner, a three-fold velocity will give a nine-fold elevation, a four-fold velocity will give a sixteen-fold elevation, and so on. The height attained, then, is not proportional to the initial velocity, but to the *square* of the velocity. As before, the work is also proportional to the weight elevated. Hence the work which any moving mass whatever is competent to perform, in virtue of the motion which it at any moment possesses, is jointly proportional to its weight and the square of its velocity. Here, then, we have a second measure of work, in which we simply translate the idea of height into its equivalent idea of motion.

In mechanics, the product of the mass of a moving body into the square of its velocity, expresses what is called the *vis viva*, or living force. It is also sometimes called the 'mechanical effect.' If, for example, a cannon pointed to the zenith urge a ball upwards with twice the velocity imparted to a second ball, the former will rise to four times the height attained by the latter. If directed against a target, it will also do four times the execution. Hence the importance of imparting a high velocity to projectiles in war. Having thus cleared our way to a perfectly definite conception of the *vis viva* of moving masses, we are prepared for the announcement that the heat generated by the shock of a falling body against the earth is proportional to the *vis viva* annihilated. The heat is proportional to the square of the velocity. In the case, therefore, of two cannon-balls of equal weight, if one strike a target with twice the velocity of the other, it will generate four times the heat, if with three times the velocity, it will generate nine times the heat, and so on.

Mr. Joule has shown that a pound weight falling from a height of 772 feet, or 772 pounds falling through one foot, will generate by its collision with the earth an amount of heat sufficient to raise a pound of water one degree Fahrenheit in temperature. 772 " foot-pounds " constitute the *mechanical equivalent* of heat. Now, a body falling from a height of 772 feet, has, upon striking the earth, a velocity of 223 feet a second; and if this velocity were imparted to the body, by any other means, the quantity of heat generated by the stoppage of its motion would be that stated above. Six times that velocity, or 1,338 feet, would not be an inordinate one for a cannon-ball as it quits the gun. Hence, a cannon-ball moving with a velocity of 1,338 feet a second, would, by collision, generate an amount of heat competent to raise its own weight of water 36 degrees Fahrenheit in temperature. If composed of iron, and if all the heat generated were concentrated in the ball itself, its temperature would be raised about 360 degrees Fahrenheit; because one degree in the case of water is equivalent to about ten degrees in the case of iron. In artillery practice, the heat generated is usually concentrated upon the front of the bolt, and on the portion of the target first struck. By this concentration the heat developed becomes sufficiently intense to raise the dust of the metal to incandescence, a flash of light often accompanying collision with the target.

Let us now fix our attention for a moment on the gunpowder which urges the cannon-ball. This is composed of combustible matter, which if burnt in the open air would yield a certain amount of heat. It will not yield this amount if it perform the work of urging a ball. The heat then generated by the gunpowder will fall short of that produced in the open air, by an amount equivalent to the *vis viva* of the ball; and this exact amount is restored by the ball on its collision with the target. In this perfect way are heat and mechanical motion connected.

Broadly enunciated, the principle of the conservation of force asserts, that the quantity of force in the universe is as unalterable as the quantity of matter; that it is alike impossible to create force and to annihilate it. But in what sense are we to understand this assertion? It would be manifestly inapplicable to the force of gravity as defined by Newton; for this is a force varying inversely as the square of the distance; and to affirm the constancy of a varying force would be self-contradictory. Yet, when the question is properly understood, gravity forms no exception to the law of conservation. Following the method pursued by Helmholtz, I will here attempt an elementary exposition of this law. Though destined in its applications to produce momentous changes in human thought, it is not difficult of comprehension.

For the sake of simplicity we will consider a particle of matter, which we may call F, to be perfectly fixed, and a second movable particle, D, placed at a distance from F. We will assume that these two particles attract each other according to the Newtonian law. At a certain distance, the attraction is of a certain definite amount, which might be determined by means of a spring balance. At half this distance the attraction would be augmented four times; at a third of the distance, nine times; at one-fourth of the distance, sixteen times, and so on. In every case, the attraction might be measured by determining, with the spring balance, the amount of tension just sufficient to prevent D from moving towards F. Thus far we have nothing whatever to do with motion; we deal with statics, not with dynamics. We simply take into account the *distance* of D from F, and the *pull* exerted by gravity at that distance.

It is customary in mechanics to represent the magnitude of a force by a line of a certain length, a force of double magnitude being represented by a line of double length, and so on. Placing then the particle D at a distance from F, we can, in imagination, draw a straight line from D to F, and at D erect a perpendicular to this line, which shall represent the amount of the attraction exerted on D. If D be at a very great distance from F, the

attraction will be very small, and the perpendicular consequently very short. If the distance be practically infinite,
the attraction is practically *nil*. Let us now suppose at
every point in the line joining ғ and ᴅ a perpendicular to
be erected, proportional in length to the attraction exerted
at that point; we thus obtain an infinite number of
perpendiculars, of gradually increasing length, as ᴅ approaches ғ. Uniting the ends of all these perpendiculars,
we obtain a curve, and between this curve and the straight
line joining ғ and ᴅ we have an area containing all the
perpendiculars placed side by side. Each one of this
infinite series of perpendiculars representing an attraction, or tension, as it is sometimes called, the area just
referred to represents the sum of the tensions exerted
upon the particle ᴅ, during its passage from its first
position to ғ.

Up to the present point we have been dealing with
tensions, not with motion. Thus far *vis viva* has been
entirely foreign to our contemplation of ᴅ and ғ. Let us
now suppose ᴅ placed at a practically infinite distance
from ғ; here, as stated, the pull of gravity would be
infinitely small, and the perpendicular representing it
would dwindle almost to a point. In this position the
sum of the tensions capable of being exerted on ᴅ would
be a maximum. Let ᴅ now begin to move in obedience
to the infinitesimal attraction exerted upon it. Motion
being once set up, the idea of *vis viva* arises. In moving
towards ғ the particle ᴅ consumes, as it were, the
tensions. Let us fix our attention on ᴅ, at any point of
the path over which it is moving. Between that point
and ғ there is a quantity of unused tensions; beyond
that point the tensions have been all consumed, but
we have in their place an equivalent quantity of *vis
viva*. After ᴅ has passed any point, the tension previously in store at that point disappears, but not with-

out having added, during the infinitely small duration of its action, a due amount of motion to that previously possessed by D. The nearer D approaches to F, the smaller is the sum of the tensions remaining, but the greater is the *vis viva*; the farther D is from F, the greater is the sum of the unconsumed tensions, and the less is the living force. Now the principle of conservation affirms *not* the constancy of the value of the tensions of gravity, nor yet the constancy of the *vis viva*, taken separately, but the absolute constancy of the value of both taken together. At the beginning the *vis viva* was zero, and the tension area was a maximum; close to F the *vis viva* is a maximum, while the tension area is zero. At every other point, the work-producing power of the particle D consists in part of *vis viva*, and in part of tensions.

If gravity, instead of being attraction, were repulsion, then, with the particles in contact, the sum of the tensions between D and F would be a maximum, and the *vis viva* zero. If, in obedience to the repulsion, D moved away from F, *vis viva* would be generated; and the farther D retreated from F the greater would be its *vis viva*, and the less the amount of tension still available for producing motion. Taking repulsion as well as attraction into account, the principle of the conservation of force affirms that the mechanical value of the *tensions* and *vires vivæ* of the material universe, so far as we know it, is a constant quantity. The universe, in short, possesses two kinds of property which are mutually convertible. The diminution of either carries with it the enhancement of the other, the total value of the property remaining unchanged.

The considerations here applied to gravity apply equally to chemical affinity. In a mixture of oxygen and hydrogen the atoms exist apart, but by the application of proper means they may be caused to rush together across that space that separates them. While this space exists, and as long as the atoms have not begun to move towards each other, we have tensions and nothing else. During their motion towards each other the tensions, as in the case of gravity, are converted into *vis viva*. After they clash we have still *vis viva*, but in another form. It *was* translation, it *is* vibration. It *was* molecular transfer, it *is* heat.

It is possible to reverse these processes, to unlock the combined atoms and replace them in their first positions. But, to accomplish this, as much heat would be required as was generated by their union. Such reversals occur daily and hourly in nature. By the solar waves, the oxygen of water is divorced from its hydrogen in the leaves of plants. As molecular *vis viva* the waves disappear, but in so doing they re-endow the atoms of oxygen and hydrogen with tension. The atoms are thus enabled to recombine, and when they do so they restore the precise amount of heat consumed in their separation. The same remarks apply to the compound of carbon and oxygen, called carbonic acid, which is exhaled from our lungs, produced by our fires, and found sparingly diffused everywhere throughout the air. In the leaves of plants the sunbeams also wrench the atoms of carbonic acid asunder, and sacrifice themselves in the act; but when the plants are burnt, the amount of heat consumed in their production is restored.

This, then, is the rhythmic play of Nature as regards her forces. Throughout all her regions she oscillates from tension to *vis viva*, from *vis viva* to tension. We have the same play in the planetary system. The earth's orbit is an ellipse, one of the foci of which is occupied by the sun. Imagine the earth at the most distant part of the orbit. Her motion, and consequently her *vis viva*, is then a minimum. The planet rounds the curve, and begins its approach to the sun. In front it has a store of tensions, which are gradually consumed, an equivalent amount of *vis viva* being generated. When nearest to the sun the motion, and consequently the *vis viva*, reach a maximum. But here the available tensions have been used up. The earth rounds this portion of the curve and retreats from the sun. Tensions are now stored up, but *vis viva* is lost, to be again restored at the expense of the complementary force on the opposite side of the curve. Thus beats the heart of the universe, but without increase or diminution of its total stock of force.

I have thus far tried to steer clear amid confusion, by fixing the mind of the reader upon things rather than upon names. But good names are essential; and here, as yet, we are not provided with such. We have had the force of gravity and living force—two utterly distinct things. We have had pulls and tensions; and we might have had the force of heat, the force of light, the force of magnetism, or the force of electricity —all of which terms have been employed more or less loosely by writers on physics. This confusion is happily avoided by the introduction of the term ' energy,' which embraces both *tension* and *vis viva*. Energy is possessed by bodies already in motion; it is then actual, and we agree to call it *actual* or *dynamic energy*. It is our old *vis viva*. On the other hand, energy is possible to bodies not in motion, but which, in virtue of attraction or repulsion, possess a power of motion which would realise itself if all hindrances were removed. Looking, for example, at gravity; a body on the earth's surface in a position from which it cannot fall to a lower one possesses no energy. It has neither

motion nor power of motion.   But the same body suspended at a height above the earth has a power of motion,
though it may not have exercised it.   Energy is possible
to such a body, and we agree to call this *potential
energy*.   It consists of our old tensions.   We, moreover, speak of the conservation of energy, instead of
the conservation of force ; and say that the sum of the
potential and dynamic energies of the material universe
is a constant quantity.

A body cast upwards consumes the actual energy of
projection, and lays up potential energy.   When it
reaches its utmost height all its actual energy is consumed, its potential energy being then a maximum.
When it returns, there is a reconversion of the potential into the actual.   A pendulum at the limit of its
swing possesses potential energy ; at the lowest point
of its arc its energy is all actual.   A patch of snow
resting on a mountain slope has potential energy ;
loosened, and shooting down as an avalanche, it possesses dynamic energy.   The pine-trees growing on the
Alps have potential energy ; but rushing down the
Holzrinne of the woodcutters they possess actual
energy.   The same is true of the mountains themselves.
As long as the rocks which compose them can fall to a
lower level, they possess potential energy, which is
converted into actual when the frost ruptures their
cohesion and hands them over to the action of gravity.
The stone avalanches of the Matterhorn and Weisshorn
are illustrations in point.   The hammer of the great
bell of Westminster, when raised before striking, possesses potential energy ; when it falls, the energy
becomes dynamic ; and after the stroke, we have the
rhythmic play of potential and dynamic in the vibrations of the bell.   The same holds good for the molecular
oscillations of a heated body.   An atom is driven

against its neighbour, and recoils. The ultimate
amplitude of the recoil being attained, the motion of
the atom in that direction is checked, and for an
instant its energy is all potential. It is then drawn
towards its neighbour with accelerated speed; thus, by
attraction, converting its potential into dynamic
energy. Its motion in this direction is also finally
checked, and again, for an instant, its energy is all
potential. It once more retreats, converting, by repulsion, its potential into dynamic energy, till the
latter attains a maximum, after which it is again
changed into potential energy. Thus, what is true of
the earth, as she swings to and fro in her yearly journey
round the sun, is also true of her minutest atom. We
have wheels within wheels, and rhythm within rhythm.

When a body is heated, a change of molecular
arrangement always occurs, and to produce this change
heat is consumed. Hence, a portion only of the heat
communicated to the body remains as dynamic energy.
Looking back on some of the statements made at the
beginning of this article, now that our knowledge is
more extensive, we see the necessity of qualifying them.
When, for example, two bodies clash, heat is generated;
but the heat, or molecular dynamic energy, developed
at the moment of collision, is not the exact equivalent
of the sensible dynamic energy destroyed. The true
equivalent is this heat, plus the potential energy conferred upon the molecules by the placing of greater·
distances between them. This molecular potential
energy is afterwards, on the cooling of the body, converted into heat.

Wherever two atoms capable of uniting together by
their mutual attractions exist separately, they form a
store of potential energy. Thus our woods, forests, and
coal-fields on the one hand, and our atmospheric oxygen

3

on the other, constitute a vast store of energy of this
kind—vast, but far from infinite. We have, besides
our coal-fields, metallic bodies more or less sparsely distributed through the earth's crust. These bodies can
be oxydised; and hence they are, so far as they go, stores
of energy. But the attractions of the great mass of the
earth's crust are already satisfied, and from them no
further energy can possibly be obtained. Ages ago the
elementary constituents of our rocks clashed together
and produced the motion of heat, which was taken up
by the ether and carried away through stellar space.
It is lost for ever as far as we are concerned. In those
ages the hot conflict of carbon, oxygen, and calcium
produced the chalk and limestone hills which are now
cold; and from this carbon, oxygen, and calcium no
further energy can be derived. So it is with almost all
the other constituents of the earth's crust. They took
their present form in obedience to molecular force; they
turned their potential energy into dynamic, and yielded
it as radiant heat to the universe, ages before man appeared upon this planet. For him a residue of potential
energy remains, vast, truly, in relation to the life and
wants of an individual, but exceedingly minute in comparison with the earth's primitive store.

To sum up. The whole stock of energy or working-
power in the world consists of attractions, repulsions,
and motions. If the attractions and repulsions be so
circumstanced as to be able to produce motion, they are
sources of working-power, but not otherwise. As stated
a moment ago, the attraction exerted between the earth
and a body at a distance from the earth's surface, is a
source of working-power; because the body can be moved
by the attraction, and in falling can perform work.
When it rests at its lowest level it is not a source of
power or energy, because it can fall no farther. But

though it has ceased to be a source of *energy*, the attraction of gravity still acts as a *force*, which holds the earth and weight together.

The same remarks apply to attracting atoms and molecules. As long as distance separates them, they can move across it in obedience to the attraction; and the motion thus produced may, by proper appliances, be caused to perform mechanical work. When, for example, two atoms of hydrogen unite with one of oxygen, to form water, the atoms are first drawn towards each other—they move, they clash, and then by virtue of their resiliency, they recoil and quiver. To this quivering motion we give the name of heat. This atomic vibration is merely the redistribution of the motion produced by the chemical affinity; and this is the only sense in which chemical affinity can be said to be converted into heat. We must not imagine the chemical attraction destroyed, or converted into anything else. For the atoms, when mutually clasped to form a molecule of water, are held together by the very attraction which first drew them towards each other. That which has really been expended is the *pull* exerted through the space by which the distance between the atoms has been diminished.

If this be understood, it will be at once seen that gravity, as before insisted on, may, in this sense, be said to be convertible into heat; that it is in reality no more an outstanding and inconvertible agent, as it is sometimes stated to be, than is chemical affinity. By the exertion of a certain pull through a certain space, a body is caused to clash with a certain definite velocity against the earth. Heat is thereby developed, and this is the only sense in which gravity can be said to be converted into heat. In no case is the *force* which produces the motion annihilated or changed into anything else.

The mutual attraction of the earth and weight exists when they are in contact, as when they were separate ; but the ability of that attraction to employ itself in the production of motion does not exist.

The transformation, in this case, is easily followed by the mind's eye.   First, the weight as a whole is set in motion by the attraction of gravity.   This motion of the mass is arrested by collision with the earth, being broken up into molecular tremors, to which we give the name of heat.

And when we reverse the process, and employ those tremors of heat to raise a weight—which is done through the intermediation of an elastic fluid in the steam-engine —a certain definite portion of the molecular motion is consumed.   In this sense, and in this sense only, can the heat be said to be converted into gravity; or, more correctly, into potential energy of gravity.   Here the destruction of the heat has created no new attraction ; but the old attraction has conferred upon it a power of exerting a certain definite pull, between the starting-point of the falling weight and the earth.

When, therefore, writers on the conservation of energy speak of tensions being 'consumed' and 'generated,' they do not mean thereby that old attractions have been annihilated, and new ones brought into existence, but that, in the one case, the power of the attraction to produce motion has been diminished by the shortening of the distance between the attracting bodies, while, in the other case, the power of producing motion has been augmented by the increase of the distance. These remarks apply to all bodies, whether they be sensible masses or molecules.

Of the inner quality that enables matter to attract matter we know nothing; and the law of conservation makes no statement regarding that quality.   It takes the facts of attraction as they stand, and affirms only the constancy of working-power. That power may exist in the form of MOTION ; or it may exist in the form of FORCE, *with distance to act through*. The former is dynamic energy, the latter is potential energy, the constancy of the sum of both being affirmed by the law of conservation. The convertibility of natural forces consists solely in transformations of dynamic into potential, and of potential into dynamic energy. In no other sense has the convertibility of force any scientific meaning.

Grave errors have been entertained as to what is really intended to be conserved by the doctrine of conservation. This exposition I hope will tend to remove them.

II.

RADIATION.[1]

1. *Visible and Invisible Radiation.*

BETWEEN the mind of man and the outer world are interposed the nerves of the human body, which translate, or enable the mind to translate, the impressions of that world into facts of consciousness and thought.

Different nerves are suited to the perception of different impressions. We do not see with the ear, nor hear with the eye, nor are we rendered sensible of sound by the nerves of the tongue. Out of the general assemblage of physical actions, each nerve, or group of nerves, selects and responds to those for the perception of which it is specially organised.

The optic nerve passes from the brain to the back of the eyeball and there spreads out, to form the retina, a web of nerve filaments, on which the images of external objects are projected by the optical portion of the eye. This nerve is limited to the apprehension of the phenomena of radiation, and, notwithstanding its marvellous sensibility to certain impressions of this class, it is singularly obtuse to other impressions.

[1] The Rede Lecture delivered in the Senate House before the University of Cambridge, May 16, 1865.

Nor does the optic nerve embrace the entire range even of radiation. Some rays, when they reach it, are incompetent to evoke its power, while others never reach it at all, being absorbed by the humours of the eye. To all rays which, whether they reach the retina or not, fail to excite vision, we give the name of invisible or obscure rays. All non-luminous bodies emit such rays. There is no body in nature absolutely cold, and every body not absolutely cold emits rays of heat. But to render radiant heat fit to affect the optic nerve a certain temperature is necessary. A cool poker thrust into a fire remains dark for a time, but when its temperature has become equal to that of the surrounding coals, it glows like them. In like manner, if a current of electricity, of gradually increasing strength, be sent through a wire of the refractory metal platinum, the wire first becomes sensibly warm to the touch; for a time its heat augments, still however remaining obscure; at length we can no longer touch the metal with impunity; and at a certain definite temperature it emits a feeble red light. As the current augments in power the light augments in brilliancy, until finally the wire appears of a dazzling white. The light which it now emits is similar to that of the sun.

By means of a prism Sir Isaac Newton unravelled the texture of solar light, and by the same simple instrument we can investigate the luminous changes of our platinum wire. In passing through the prism all its rays (and they are infinite in variety) are bent or refracted from their straight course; and, as different rays are differently refracted by the prism, we are by it enabled to separate one class of rays from another. By such prismatic analysis Dr. Draper has shown, that when the platinum wire first begins to glow, the light emitted is sensibly red. As the glow augments the red becomes more brilliant, but at the same time orange rays are added to the emission. Augmenting the temperature still further, yellow rays appear beside the orange; after the yellow, green rays are emitted ; and after the green come, in succession, blue, indigo, and violet rays. To display all these colours at the same time the platinum wire must be *white-hot*: the impression of whiteness being in fact produced by the simultaneous action of all these colours on the optic nerve.

In the experiment just described we began with a platinum wire at an ordinary temperature, and gradually raised it to a white heat. At the beginning, and even before the electric current had acted at all upon the wire, it emitted invisible rays. For some time after the action of the current had commenced, and even for a time after the wire had become intolerable to the touch, its radiation was still invisible. The question now arises, What becomes of these invisible rays when the visible ones make their appearance ? It will be proved in the sequel that they maintain themselves in the radiation ; that a ray once emitted continues to be emitted when the temperature is increased, and hence the emission from our platinum wire, even when it has attained its maximum brilliancy, consists of a mixture of visible and invisible rays. If, instead of the platinum wire, the earth itself were raised to incandescence, the obscure radiation which it now emits would continue to be emitted. To reach incandescence the planet would have to pass through all the stages of non-luminous radiation, and the final emission would embrace the rays of all these stages. There can hardly be a doubt that from the sun itself, rays proceed similar in kind to those which the dark earth pours nightly into space. In fact, the various kind of obscure rays emitted by all the planets of our system are included in the present radiation of the sun.

The great pioneer in this domain of science was Sir William Herschel. Causing a beam of solar light to pass through a prism, he resolved it into its coloured constituents; he formed what is technically called the solar spectrum. Exposing thermometers to the successive colours he determined their heating power, and found it to augment from the violet or most refracted end, to the red or least refracted end of the spectrum. But he did not stop here. Pushing his thermometers into the dark space beyond the red he found that, though the light had disappeared, the radiant heat falling on the instruments was more intense than that at any visible part of the spectrum. In fact, Sir William Herschel showed, and his results have been verified by various philosophers since his time, that, besides its luminous rays, the sun pours forth a multitude of other rays, more powerfully calorific than the luminous ones, but entirely unsuited to the purposes of vision.

At the less refrangible end of the solar spectrum, then, the range of the sun's radiation is not limited by that of the eye. The same statement applies to the more refrangible end. Ritter discovered the extension of the spectrum into the invisible region beyond the violet; and, in recent times, this ultra-violet emission has had peculiar interest conferred upon it by the admirable researches of Professor Stokes. The complete spectrum of the sun consists, therefore, of three distinct parts :—first, of ultra-red rays of high heating power, but unsuited to the purposes of vision; secondly, of luminous rays which display the succession of colours, red, orange, yellow, green, blue, indigo, violet; thirdly, of ultra-violet rays which, like the ultra-red ones, are incompetent to excite vision, but which, unlike the ultra-red rays, possess a very feeble heating power. In consequence, however, of their chemical energy these ultra-violet rays are of the utmost importance to the organic world.

2. *Origin and Character of Radiation. The Ether.*

When we see a platinum wire raised gradually to a white heat, and emitting in succession all the colours of the spectrum, we are simply conscious of a series of changes in the condition of our own eyes. We do not see the actions in which these successive colours originate, but the mind irresistibly infers that the appearance of the colours corresponds to certain contemporaneous changes in the wire. What is the nature of these changes? In virtue of what condition does the wire radiate at all? We must now look from the wire, as a whole, to its constituent atoms. Could we see those atoms, even before the electric current has begun to act upon them, we should find them in a state of vibration. In this vibration, indeed, consists such warmth as the wire then possesses. Locke enunciated this idea with great precision, and it has been placed beyond the pale of doubt by the excellent quantitative researches of Mr. Joule. ' Heat,' says Locke, ' is a very brisk agitation of the insensible parts of the object, which produce in us that sensation from which we denominate the object hot: so what in our sensations is *heat* in the object is nothing but *motion*.' When the electric current, still feeble, begins to pass through the wire, its first act is to intensify the vibrations already existing, by causing the atoms to swing through wider ranges. Technically speaking, the *amplitudes* of the oscillations are increased. The current does this, however, without altering the periods of the old vibrations, or the times in which they were executed. But besides intensifying the old vibrations the current generates new and more rapid ones, and when a certain efinite rapidity has been attained, the wire begins to glow. The colour first exhibited is red, which corresponds to the lowest rate of vibration of which the eye is able to take cognisance. By augmenting the strength of the electric current more rapid vibrations are introduced, and orange rays appear. A quicker rate of vibration produces yellow, a still quicker green; and by further augmenting the rapidity, we pass through blue, indigo, and violet, to the extreme ultra-violet rays.

Such are the changes recognised by the mind in the wire itself, as concurrent with the visual changes taking place in the eye. But what connects the wire with this organ? By what means does it send such intelligence of its varying condition to the optic nerve? Heat being as defined by Locke, ' a very brisk agitation of the insensible parts of an object,' it is readily conceivable that on touching a heated body the agitation may communicate itself to the adjacent nerves, and announce itself to them as light or heat. But the optic nerve does not touch the hot platinum, and hence the pertinence of the question, By what agency are the vibrations of the wire transmitted to the eye?

The answer to this question involves one of the most important physical conceptions that the mind of man has yet achieved: the conception of a medium filling space and fitted mechanically for the transmission of the vibrations of light and heat, as air is fitted for the transmission of sound. This medium is called the *luminiferous ether*. Every vibration of every atom of our platinum wire raises in this ether a wave, which speeds through it at the rate of 186,000 miles a second.

The ether suffers no rupture of continuity at the surface of the eye, the inter-molecular spaces of the various humours are filled with it; hence the waves generated by the glowing platinum can cross these humours and impinge on the optic nerve at the back of the eye.[1] Thus the sensation of light reduces itself to the acceptance of motion. Up to this point we deal with pure mechanics; but the subsequent translation of the shock of the ethereal waves into consciousness eludes mechanical science. As an oar dipping into the Cam generates systems of waves, which, speeding from the centre of disturbance, finally stir the sedges on the river's bank, so do the vibrating atoms generate in the surrounding ether undulations, which finally stir the filaments of the retina. The motion thus imparted is transmitted with measurable, and not very great velocity to the brain, where, by a process which the science of mechanics does not even tend to unravel, the tremor of the nervous matter is converted into the conscious impression of light.

Darkness might then be defined as ether at rest; light as ether in motion. But in reality the ether is never at rest, for in the absence of light-waves we have heat-waves always speeding through it. In the spaces of the universe both classes of undulations incessantly commingle. Here the waves issuing from uncounted centres cross, coincide, oppose, and pass through each other, without confusion or ultimate extinction. Every star is seen across the entanglement of wave-motions produced by all other stars. It is the ceaseless thrill caused by those distant orbs collectively in the ether, that constitutes what we call the ' temperature of space.' As the air of a room accommodates itself to the require-

---

[1] The action here described is analogous to the passage of sound-waves through thick felt whose interstices are occupied by air.

ments of an orchestra, transmitting each vibration of
every pipe and string, so does the inter-stellar ether
accommodate itself to the requirements of light and
heat. Its waves mingle in space without disorder,
each being endowed with an individuality as indestructible as if it alone had disturbed the universal
repose.

All vagueness with regard to the use of the terms
'radiation' and 'absorption' will now disappear.
Radiation is the communication of vibratory motion to
the ether; and when a body is said to be chilled by
radiation, as for example the grass of a meadow on a
starlight night, the meaning is, that the molecules of
the grass have lost a portion of their motion, by imparting it to the medium in which they vibrate. On
the other hand, the waves of ether may so strike
against the molecules of a body exposed to their action
as to yield up their motion to the latter; and in this
transfer of the motion from the ether to the molecules
consists the absorption of radiant heat. All the phenomena of heat are in this way reducible to interchanges
of motion; and it is purely as the recipients or the
donors of this motion, that we ourselves become conscious of the action of heat and cold.

3. *The Atomic Theory in reference to the Ether.*

The word 'atoms' has been more than once employed in this discourse. Chemists have taught us that
all matter is reducible to certain elementary forms to
which they give this name. These atoms are endowed
with powers of mutual attraction, and under suitable
circumstances they coalesce to form compounds. Thus
oxygen and hydrogen are elements when separate, or

merely *mixed*, but they may be made to *combine* so as to form molecules, each consisting of two atoms of hydrogen and one of oxygen. In this condition they constitute water. So also chlorine and sodium are elements, the former a pungent gas, the latter a soft metal; and they unite together to form chloride of sodium or common salt. In the same way the element nitrogen combines with hydrogen, in the proportion of one atom of the former to three of the latter, to form ammonia. Picturing in imagination the atoms of elementary bodies as little spheres, the molecules of compound bodies must be pictured as groups of such spheres. This is the atomic theory as Dalton conceived it. Now if this theory have any foundation in fact, and if the theory of an ether pervading space, and constituting the vehicle of atomic motion, be founded in fact, it is surely of interest to examine whether the vibrations of elementary bodies are modified by the act of combination—whether as regards radiation and absorption, or, in other words, whether as regards the communication of motion to the ether, and the acceptance of motion from it, the deportment of the uncombined atoms will be different from that of the combined.

4. *Absorption of Radiant Heat by Gases.*

We have now to submit these considerations to the only test by which they can be tried, namely, that of experiment. An experiment is well defined as a question put to Nature; but, to avoid the risk of asking amiss, we ought to purify the question from all adjuncts which do not necessarily belong to it. Matter has been shown to be composed of elementary constituents, by the compounding of which all its varieties are produced. But, besides the chemical unions which they form, both elementary and compound bodies can unite in another and less intimate way. Gases and vapours aggregate to liquids and solids, without any change of their chemical nature. We do not yet know how the transmission of radiant heat may be affected by the entanglement due to cohesion ; and, as our object now is to examine the influence of chemical union alone, we shall render our experiments more pure by liberating the atoms and molecules entirely from the bonds of cohesion, and employing them in the gaseous or vaporous form.

Let us endeavour to obtain a perfectly clear mental image of the problem now before us. Limiting in the first place our enquiries to the phenomena of absorption, we have to picture a succession of waves issuing from a radiant source and passing through a gas ; some of them striking against the gaseous molecules and yielding up their motion to the latter; others gliding round the molecules, or passing through the intermolecular spaces without apparent hindrance. The problem before us is to determine whether such free molecules have any power whatever to stop the waves of heat; and if so, whether different molecules possess this power in different degrees.

In examining the problem let us fall back upon an actual piece of work, choosing as the source of our heat-waves a plate of copper, against the back of which a steady sheet of flame is permitted to play. On emerging from the copper, the waves, in the first instance, pass through a space devoid of air, and then enter a hollow glass cylinder, three feet long and three inches wide. The two ends of this cylinder are stopped by two plates of rock-salt, a solid substance which offers a scarcely sensible obstacle to the passage of the calorific waves. After passing through the tube, the radiant heat falls upon the anterior face of a thermo-electric pile,[1] which instantly converts the heat into an electric current. This current conducted round a magnetic needle deflects it, and the magnitude of the deflection is a measure of the heat falling upon the pile. This famous instrument, and not an ordinary thermometer, is what we shall use in these enquiries, but we shall use it in a somewhat novel way. As long as the two opposite faces of the thermo-electric pile are kept at the same temperature, no matter how high that may be, there is no current generated. The current is a consequence of a difference of temperature between the two opposite faces of the pile. Hence, if after the anterior face has received the heat from our radiating source, a second source, which we may call the compensating source, be permitted to radiate against the posterior face, this latter radiation will tend to neutralise the former. When the neutralisation is perfect, the magnetic needle connected with the pile is no longer deflected, but points to the zero of the graduated circle over which it hangs.

And now let us suppose the glass tube, through which the waves from the heated plate of copper are passing, to be exhausted by an air-pump, the two sources of heat acting at the same time on the two opposite faces of the pile. When by means of an adjusting screen, perfectly equal quantities of heat are imparted to the two faces, the needle points to zero. Let any gas be now permitted to enter the exhausted tube; if its molecules possess any power of intercepting the calorific waves, the equilibrium previously existing will be destroyed, the compensating source will triumph,

[1] In the Appendix to the first chapter of 'Heat as a Mode of Motion,' the construction of the thermo-electric pile is fully explained.

and a deflection of the magnetic needle will be the immediate consequence. From the deflections thus produced by different gases, we can readily deduce the relative amounts of wave-motion which their molecules intercept.

In this way the substances mentioned in the following table were examined, a small portion only of each being admitted into the glass tube. The quantity admitted in each case was just sufficient to depress a column of mercury associated with the tube one inch : in other words, the gases were examined at a pressure of one-thirtieth of an atmosphere. The numbers in the table express the relative amounts of wave-motion absorbed by the respective gases, the quantity intercepted by atmospheric air being taken as unity.

Radiation through Gases.

Name of gas	Relative absorption
Air	1
Oxygen	1
Nitrogen	1
Hydrogen	1
Carbonic oxide	750
Carbonic acid	972
Hydrochloric acid	1,005
Nitric oxide	1,590
Nitrous oxide	1,860
Sulphide of hydrogen	2,100
Ammonia	5,460
Olefiant gas	6,030
Sulphurous acid	6,480

Every gas in this table is perfectly transparent to light, that is to say, all waves within the limits of the visible spectrum pass through it without obstruction ; but for the waves of slower period, emanating from our heated plate of copper, enormous differences of absorptive power are manifested. These differences illustrate

4

in the most unexpected manner the influence of chemical combination. Thus the elementary gases, oxygen, hydrogen, and nitrogen, and the mixture atmospheric air, prove to be practical vacua to the rays of heat; for every ray, or, more strictly speaking, for every unit of wave-motion, which any one of them intercepts, perfectly transparent ammonia intercepts 5,460 units, olefiant gas 6,030 units, while sulphurous acid gas absorbs 6,480 units. What becomes of the wave-motion thus intercepted? It is applied to the heating of the absorbing gas. Through air, oxygen, hydrogen, and nitrogen, the waves of ether pass without absorption, and these gases are not sensibly changed in temperature by the most powerful calorific rays. The position of nitrous oxide in the foregoing table is worthy of particular notice. In this gas we have the same atoms in a state of chemical union, that exist uncombined in the atmosphere; but the absorption of the compound is 1,800 times that of air.

5. *Formation of Invisible Foci.*

This extraordinary deportment of the elementary gases naturally directed attention to elementary bodies in other states of aggregation. Some of Melloni's results now attained a new significance. This celebrated experimenter had found crystals of sulphur to be highly pervious to radiant heat; he had also proved that lamp-black, and black glass, (which owes its blackness to the element carbon) were to a considerable extent transparent to calorific rays of low refrangibility. These facts, harmonising so strikingly with the deportment of the simple gases, suggested further enquiry. Sulphur dissolved in bisulphide of carbon was found almost perfectly diathermic. The dense and deeply-coloured element bromine was examined, and found competent to cut off the light of our most brilliant flames, while it transmitted the invisible calorific rays with extreme freedom. Iodine, the companion element of bromine, was next thought of, but it was found impracticable to examine the substance in its usual solid condition. It however dissolves freely in bisulphide of carbon. There is no chemical union between the liquid and the iodine; it is simply a case of solution, in which the uncombined atoms of the element can act upon the radiant heat. When permitted to do so, it was found that a layer of dissolved iodine, sufficiently opaque to cut off the light of the midday sun, was almost absolutely transparent to the invisible calorific rays.[1]

By prismatic analysis Sir William Herschel separated the luminous from the non-luminous rays of the sun, and he also sought to render the obscure rays visible by concentration. Intercepting the luminous portion of his spectrum he brought, by a converging lens, the ultra-red rays to a focus, but by this condensation he obtained no light. The solution of iodine offers a means of filtering the solar beam, or failing it, the beam of the electric lamp, which renders attainable far more powerful foci of invisible rays than could possibly be obtained by the method of Sir William Herschel. For to form his spectrum he was obliged to operate upon solar light which had passed through a narrow slit or through a small aperture, the amount of the obscure heat being limited by this circumstance. But with our opaque solution we may employ the entire surface of the largest lens, and having thus converged the rays,

---

[1] Professor Dewar has recently succeeded in producing a medium highly opaque to light, and highly transparent to obscure heat, by fusing together sulphur and iodine.

luminous and non-luminous, we can intercept the former by the iodine, and do what we please with the latter. Experiments of this character, not only with the iodine solution, but also with black glass and layers of lamp-black, were publicly performed at the Royal Institution in the early part of 1862, and the effects at the foci of invisible rays, then obtained, were such as had never been witnessed previously.

In the experiments here referred to, glass lenses were employed to concentrate the rays. But glass, though highly transparent to the luminous, is in a high degree opaque to the invisible, heat-rays of the electric lamp, and hence a large portion of those rays was intercepted by the glass. The obvious remedy here is to employ rock-salt lenses instead of glass ones, or to abandon the use of lenses wholly, and to concentrate the rays by a metallic mirror. Both of these improvements have been introduced, and, as anticipated, the invisible foci have been thereby rendered more intense. The mode of operating remains however the same, in principle, as that made known in 1862. It was then found that an instant's exposure of the face of the thermo-electric pile to the focus of invisible rays, dashed the needles of a coarse galvanometer violently aside. It is now found that on substituting for the face of the thermo-electric pile a combustible body, the invisible rays are competent to set that body on fire.

6. *Visible and Invisible Rays of the Electric Light.*

We have next to examine what proportion the non-luminous rays of the electric light bear to the luminous ones. This the opaque solution of iodine enables us to do with an extremely close approximation to the truth.

The pure bisulphide of carbon, which is the solvent of the iodine, is perfectly transparent to the luminous, and almost perfectly transparent to the dark, rays of the electric lamp. Supposing the total radiation of the lamp to pass through the transparent bisulphide, while through the solution of iodine only the dark rays are transmitted. If we determine, by means of a thermo-electric pile, the total radiation, and deduct from it the purely obscure, we obtain the value of the purely luminous emission. Experiments, performed in this way, prove that if all the visible rays of the electric light were converged to a focus of dazzling brilliancy, its heat would only be one-eighth of that produced at the unseen focus of the invisible rays.

Exposing his thermometers to the successive colours of the solar spectrum, Sir William Herschel determined the heating power of each, and also that of the region beyond the extreme red. Then drawing a straight line to represent the length of the spectrum, he erected, at various points, perpendiculars to represent the calorific intensity existing at those points. Uniting the ends of all his perpendiculars, he obtained a curve which showed at a glance the manner in which the heat was distributed in the solar spectrum. Professor Müller of Freiburg, with improved instruments, afterwards made similar experiments, and constructed a more accurate diagram of the same kind. We have now to examine the distribution of heat in the spectrum of the electric light; and for this purpose we shall employ a particular form of the thermo-electric pile, devised by Melloni. Its face is a rectangle, which by means of movable side-pieces can be rendered as narrow as desired. We can, for example, have the face of the pile the tenth, the hundredth, or even the thousandth of an inch in breadth. By means of an endless screw, this linear thermo-electric pile may be moved through the entire spectrum, from the violet to the red, the amount of heat falling upon the pile at every point of its march, being declared by a magnetic needle associated with the pile.

When this instrument is brought up to the violet end of the spectrum of the electric light, the heat is found to be insensible. As the pile is gradually moved from the violet end towards the red, heat soon manifests itself, augmenting as we approach the red. Of all the colours of the visible spectrum the red possesses the highest heating power. On pushing the pile into the dark region beyond the red, the heat, instead of vanishing, rises suddenly and enormously in intensity, until at some distance beyond the red it attains a maximum. Moving the pile still forward, the thermal power falls, somewhat more rapidly than it rose. It then gradually shades away, but, for a distance beyond the red greater than the length of the whole visible spectrum, signs of heat may be detected.

Drawing a datum line, and erecting along it perpendiculars, proportional in length to the thermal intensity at the respective points, we obtain the extraordinary curve, shown on the opposite page, which exhibits the distribution of heat in the spectrum of the electric light. In the region of dark rays, beyond the red, the curve shoots up to B, in a steep and massive peak—a kind of Matterhorn of heat, which dwarfs the portion of the diagram C D E, representing the luminous radiation. Indeed the idea forced upon the mind by this diagram is that the light rays are a mere insignificant appendage to the heat-rays represented by the area A B C D, thrown in as it were by nature for the purpose of vision.

The diagram drawn by Professor Müller to represent

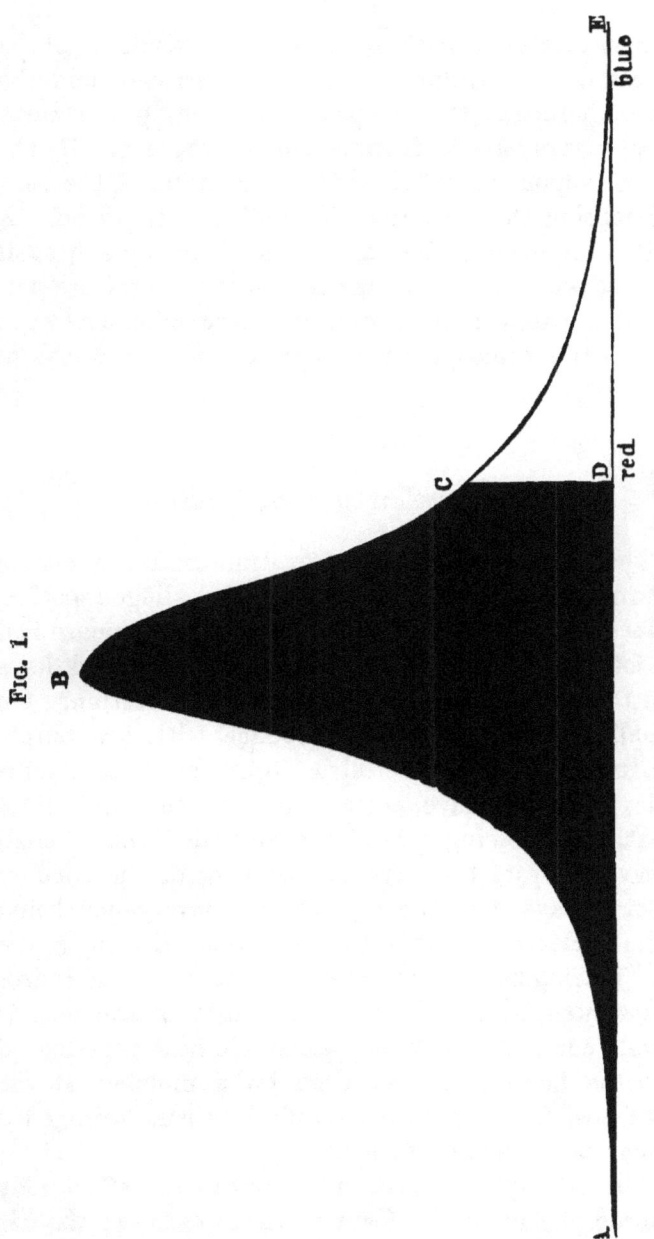

Fig. 1.

SPECTRUM OF ELECTRIC LIGHT.

the distribution of heat in the solar spectrum is not by any means so striking as that just described, and the reason, doubtless, is that prior to reaching the earth the solar rays have to traverse our atmosphere. By the aqueous vapour there diffused, the summit of the peak representing the sun's invisible radiation is cut off. A similar lowering of the mountain of invisible heat is observed when the rays from the electric light are permitted to pass through a film of water, which acts upon them as the atmospheric vapour acts upon the rays of the sun.

7. *Combustion by Invisible Rays.*

The sun's invisible rays far transcend the visible ones in heating power, so that if the alleged performances of Archimedes during the siege of Syracuse had any foundation in fact, the dark solar rays would have been the philosopher's chief agents of combustion. On a small scale we can readily produce, with the purely invisible rays of the electric light, all that Archimedes is said to have performed with the sun's total radiation. Placing behind the electric light a small concave mirror, the rays are converged, the cone of reflected rays and their point of convergence being rendered clearly visible by the dust always floating in the air. Placing between the luminous focus and the source of rays our solution of iodine, the light of the cone is entirely cut away; but the intolerable heat experienced when the hand is placed, even for a moment, at the dark focus, shows that the calorific rays pass unimpeded through the opaque solution.

Almost anything that ordinary fire can effect may be accomplished at the focus of invisible rays; the *air* at the focus remaining at the same time perfectly cold, on account of its transparency to the heat-rays. An air thermometer, with a hollow rock-salt bulb, would be unaffected by the heat of the focus: there would be no expansion, and in the open air there is no convection. The ether at the focus, and not the air, is the substance in which the heat is embodied. A block of wood, placed at the focus, absorbs the heat, and dense volumes of smoke rise swiftly upwards, showing the manner in which the air itself would rise, if the invisible rays were competent to heat it. At the perfectly dark focus dry paper is instantly inflamed: chips of wood are speedily burnt up: lead, tin, and zinc are fused: and disks of charred paper are raised to vivid incandescence. It might be supposed that the obscure rays would show no preference for black over white; but they do show a preference, and to obtain rapid combustion, the body, if not already black, ought to be blackened. When metals are to be burned, it is necessary to blacken or otherwise tarnish them, so as to diminish their reflective power. Blackened zinc foil, when brought into the focus of invisible rays, is instantly caused to blaze, and burns with its peculiar purple light. Magnesium wire flattened, or tarnished magnesium ribbon, also bursts into flame. Pieces of charcoal suspended in a receiver full of oxygen are also set on fire when the invisible focus falls upon them; the dark rays after having passed through the receiver, still possessing sufficient power to ignite the charcoal, and thus initiate the attack of the oxygen. If, instead of being plunged in oxygen, the charcoal be suspended in vacuo, it immediately glows at the place where the focus falls.

8. *Transmutation of Rays :* [1] *Calorescence.*

Eminent experimenters were long occupied in demonstrating the substantial identity of light and radiant heat, and we have now the means of offering a new and striking proof of this identity. A concave mirror produces, beyond the object which it reflects, an inverted and magnified image of the object. Withdrawing, for example, our iodine solution, an intensely luminous inverted image of the carbon points of the electric light is formed at the focus of the mirror employed in the foregoing experiments. When the solution is interposed, and the light is cut away, what becomes of this image? It disappears from sight; but an invisible thermograph remains, and it is only the peculiar constitution of our eyes that disqualifies us from seeing the picture formed by the calorific rays. Falling on white paper, the image chars itself out: falling on black paper, two holes are pierced in it, corresponding to the images of the two coke points: but falling on a thin plate of carbon in vacuo, or upon a thin sheet of platinised platinum, either in vacuo or in air, radiant heat is converted into light, and the image stamps itself in vivid incandescence upon both the carbon and the metal. Results similar to those obtained with the electric light have also been obtained with the invisible rays of the lime-light and of the sun.

Before a Cambridge audience it is hardly necessary to refer to the excellent researches of Professor Stokes at the opposite end of the spectrum. The above results constitute a kind of complement to his discoveries. Professor Stokes named the phenomena which he has

[1] I borrow this term from Professor Challis, 'Philosophical Magazine,' vol. xii. p. 521.

discovered and investigated *Fluorescence*; for the new phenomena here described I have proposed the term *Calorescence*. He, by the interposition of a proper medium, so lowered the refrangibility of the ultra-violet rays of the spectrum as to render them visible. Here, by the interposition of the platinum foil, the refrangibility of the ultra-red rays is so exalted as to render them visible. Looking through a prism at the incandescent image of the carbon points, the light of the image is decomposed, and a complete spectrum is obtained. The invisible rays of the electric light, remoulded by the atoms of the platinum, shine thus visibly forth; ultra-red rays being converted into red, orange, yellow, green, blue, indigo, violet, and ultra-violet ones. Could we, moreover, raise the original source of rays to a sufficiently high temperature, we might not only obtain from the dark rays of such a source a single incandescent image, but from the dark rays of this image we might obtain a second one, from the dark rays of the second a third, and so on—a series of complete images and spectra being thus extracted from the invisible emission of the primitive source.[1]

[1] On investigating the calorescence produced by rays transmitted through glasses of various colours, it was found that in the case of certain specimens of blue glass, the platinum foil glowed with a *pink* or *purplish* light. The effect was not subjective, and consider-ations of obvious interest are suggested by it. Different kinds of black glass differ notably as to their power of transmitting radiant heat. When thin, some descriptions tint the sun with a greenish hue : others make it appear a glowing red without any trace of green. The latter are far more diathermic than the former. In fact, carbon when perfectly dissolved and incorporated with a good white glass, is highly transparent to the calorific rays, and by em-ploying it as an absorbent the phenomena of 'calorescence' may be obtained, though in a less striking form than with the iodine. The black glass chosen for thermometers, and intended to absorb com-pletely the solar heat, may entirely fail in this object, if the glass in which the carbon is incorporated be colourless. To render the

9. *Deadness of the Optic Nerve to the Calorific Rays.*

The layer of iodine used in the foregoing experiments intercepted the rays of the noonday sun. No trace of light from the electric lamp was visible in the darkest room, even when a white screen was placed at the focus of the mirror employed to concentrate the light. It was thought, however, that if the retina itself were brought into the focus the sensation of light might be experienced. The danger of this experiment was twofold. If the dark rays were absorbed in a high degree by the humours of the eye the albumen of the humours might coagulate along the line of the rays. If, on the contrary, no such high absorption took place, the rays might reach the retina with a force sufficient to destroy it. To test the likelihood of these results, experiments were made on water and on a solution of alum, and they showed it to be very improbable that in the brief time requisite for an experiment any serious damage could be done. The eye was therefore caused to approach the dark focus, no defence, in the first instance, being provided; but the heat, acting upon the parts surrounding the pupil, could not be borne. An aperture was therefore pierced in a plate of metal, and the eye, placed behind the aperture, was caused to approach the point of convergence of invisible rays. The focus was attained, first by the pupil and afterwards by the retina. Removing the eye, but permitting the plate of metal to remain, a sheet of platinum foil was placed in the position occupied by the

bulb of a thermometer a perfect absorbent, the glass ought in the first instance to be green. Soon after the discovery of fluorescence the late Dr. William Allen Miller pointed to the lime-light as an illustration of exalted refrangibility. Direct experiments have since entirely confirmed the view expressed at page 210 of his work on ' Chemistry,' published in 1855.

retina a moment before. The platinum became red-hot. No sensible damage was done to the eye by this experiment; no impression of light was produced; the optic nerve was not even conscious of heat.

But the humours of the eye are known to be highly impervious to the invisible calorific rays, and the question therefore arises, 'Did the radiation in the foregoing experiment reach the retina at all?' The answer is, that the rays were in part transmitted to the retina, and in part absorbed by the humours. Experiments on the eye of an ox showed that the proportion of obscure rays which reached the retina amounted to 18 per cent. of the total radiation; while the luminous emission from the electric light amounts to no more than 10 per cent. of the same total. Were the purely luminous rays of the electric lamp converged by our mirror to a focus, there can be no doubt as to the fate of a retina placed there. Its ruin would be inevitable; and yet this would be accomplished by an amount of wave-motion but little more than half of that which the retina, without exciting consciousness, bears at the focus of invisible rays.

This subject will repay a moment's further attention. At a common distance of a foot the visible radiation of the electric light employed in these experiments is 800 times the light of a candle. At the same distance, the portion of the radiation of the electric light which reaches the retina, but fails to excite vision, is about 1,500 times the luminous radiation of the candle.[1] But a candle on a clear night can readily be seen at a distance of a mile, its light at this distance being less than $\frac{1}{20,000,000}$ of its light at the distance of a foot.

[1] It will be borne in mind that the heat which any ray, luminous or non-luminous, is competent to generate is the true measure of the energy of the ray.

Hence, to make the candle-light a mile off equal in
power to the non-luminous radiation received from the
electric light at a foot distance, its intensity would
have to be multiplied by $1,500 \times 20,000,000$, or by
thirty thousand millions. Thus the thirty thousand
millionth part of the invisible radiation from the
electric light, received by the retina at the distance of
a foot, would, if slightly changed in character, be amply
sufficient to provoke vision. Nothing could more
forcibly illustrate that special relationship supposed by
Melloni and others to subsist between the optic nerve
and the oscillating periods of luminous bodies. The
optic nerve responds, as it were, to the waves with
which it is in consonance, while it refuses to be excited
by others of almost infinitely greater energy, whose
periods of recurrence are not in unison with its own.

10. *Persistence of Rays.*

At an early part of this lecture it was affirmed, that
when a platinum wire was gradually raised to a state
of high incandescence, new rays were constantly added,
while the intensity of the old ones was increased. Thus,
in Dr. Draper's experiments, the rise of temperature
that generated the orange, yellow, green, and blue
augmented the intensity of the red. What is true of
the red is true of every other ray of the spectrum,
visible and invisible. We cannot indeed *see* the augmentation of intensity in the region beyond the red,
but we can measure it and express it numerically.
With this view the following experiment was performed:
A spiral of platinum wire was surrounded by a small
glass globe to protect it from currents of air; through
an orifice in the globe the rays could pass from the

spiral and fall afterwards upon a thermo-electric pile. Placing in front of the orifice an opaque solution of iodine, the platinum was gradually raised from a low dark heat to the fullest incandescence, with the following results:—

Appearance of spiral	Energy of obscure radiation
Dark	1
Dark, but hotter	3
Dark, but still hotter	5
Dark, but still hotter	10
Feeble red	19
Dull red	25
Red	37
Full red	62
Orange	89
Bright orange	144
Yellow	202
White	276
Intense white	440

Thus the augmentation of the electric current, which raises the wire from its primitive dark condition to an intense white heat, exalts at the same time the energy of the obscure radiation, until at the end it is fully 440 times what it was at the beginning.

What has been here proved true of the totality of the ultra-red rays is true for each of them singly. Placing our linear thermo-electric pile in any part of the ultra-red spectrum, it may be proved that a ray once emitted continues to be emitted with increased energy as the temperature is augmented. The platinum spiral, so often referred to, being raised to whiteness by an electric current, a brilliant spectrum was formed from its light. A linear thermo-electric pile was placed in the region of obscure rays beyond the red, and by diminishing the current the spiral was reduced to a low temperature. It was then caused to pass through

various degrees of darkness and incandescence, with
the following results:—

Appearance of spiral						Energy of obscure rays
Dark	.	.	.	.	.	1
Dark	.	.	.	.	.	6
Faint red	.	.	.	.	.	10
Dull red	.	.	.	.	.	13
Red	.	.	.	.	.	18
Full red.	.	.	.	.	.	27
Orange .	.	.	.	.	.	60
Yellow .	.	.	.	.	.	93
White .	.	.	.	.	.	122

Here, as in the former case, the dark and bright
radiations reached their maximum together; as the
one augmented, the other augmented, until at last the
energy of the obscure rays of the particular refrangibility here chosen, became 122 times what it was at
first. To reach a white heat the wire has to pass
through all the stages of invisible radiation, but in its
most brilliant condition it embraces, in an intensified
form, the rays of all those stages.

And thus it is with all other kinds of matter, as far
as they have hitherto been examined. Coke, whether
brought to a white heat by the electric current, or by
the oxyhydrogen jet, pours out invisible rays with
augmented energy, as its light is increased. The same
is true of lime, bricks, and other substances. It is
true of all metals which are capable of being heated to
incandescence. It also holds good for phosphorus
burning in oxygen. Every gush of dazzling light has
associated with it a gush of invisible radiant heat,
which far transcends the light in energy. This condition of things applies to all bodies capable of being
raised to a white heat, either in the solid or the molten
condition. It would doubtless also apply to the

luminous fogs formed by the condensation of incandescent vapours. In such cases when the curve representing the radiant energy of the body is constructed, the obscure radiation towers upwards like a mountain, the luminous radiation resembling a mere ' spur ' at its base. From the very brightness of the light of some of the fixed stars we may infer the intensity of that dark radiation, which is the precursor and inseparable associate of their luminous rays.

We thus find the luminous radiation appearing when the radiant body has attained a certain temperature; or, in other words, when the vibrating atoms of the body have attained a certain width of swing. In solid and molten bodies a certain amplitude cannot be surpassed without the introduction of periods of vibration, which provoke the sense of vision. How are we to figure this? If permitted to speculate, we might ask, are not these more rapid vibrations the progeny of the slower? Is it not really the mutual action of the atoms, when they swing through very wide spaces, and thus encroach upon each other, that causes them to tremble in quicker periods? If so, whatever be the agency by which the large swinging space is obtained, we shall have light-giving vibrations associated with it. It matters not whether the large amplitudes be produced by the strokes of a hammer, or by the blows of the molecules of a non-luminous gas, like air at some height above a gas-flame; or by the shock of the ether particles when transmitting radiant heat. The result in all cases will be incandescence. Thus, the invisible waves of our filtered electric beam may be regarded as generating synchronous vibrations among the atoms of the platinum on which they impinge; but, once these vibrations have attained a certain amplitude, the mutual jostling of the atoms produces quicker tremors,

and the light-giving waves follow as the necessary
product of the heat-giving ones.

.

11. *Absorption of Radiant Heat by Vapours and Odours.*

We commenced the demonstrations brought forward
in this lecture by experiments on permanent gases, and
we have now to turn our attention to the vapours of
volatile liquids. Here, as in the case of the gases,
vast differences have been proved to exist between
various kinds of molecules, as regards their power of
intercepting the calorific waves. While some vapours
allow the waves a comparatively free passage, the
faintest mixture of other vapours causes a deflection of
the magnetic needle. Assuming the absorption effected
by air, at a pressure of one atmosphere, to be unity,
the following are the absorptions effected by a series of
vapours at a pressure of $\frac{1}{60}$th of an atmosphere :—

Name of vapour				Absorption
Bisulphide of carbon	.	.	.	47
Iodide of methyl .	.	.	.	115
Benzol .	.	.	.	136
Amylene	.	.	.	321
Sulphuric ether	.	.	.	440
Formic ether.	.	.	.	548
Acetic ether .	.	.	.	612

Bisulphide of carbon is the most transparent vapour
in this list ; and acetic ether the most opaque ; $\frac{1}{60}$th of
an atmosphere of the former, however, produces 47
times the effect of a whole atmosphere of air, while $\frac{1}{60}$th
of an atmosphere of the latter produces 612 times the
effect of a whole atmosphere of air. Reducing dry air
to the pressure of the acetic ether here employed, and
comparing them then together, the quantity of wave-

motion intercepted by the ether would be many thousand times that intercepted by the air.

Any one of these vapours discharged into the free atmosphere, in front of a body emitting obscure rays, intercepts more or less of the radiation. A similar effect is produced by perfumes diffused in the air, though their attenuation is known to be almost infinite. Carrying, for example, a current of dry air over bibulous paper, moistened by patchouli, the scent taken up by the current absorbs 30 times the quantity of heat intercepted by the air which carries it; and yet patchouli acts more feebly on radiant heat than any other perfume yet examined. Here follow the results obtained with various essential oils, the odour, in each case, being carried by a current of dry air into the tube already employed for gases and vapours :—

Name of perfume	Absorption
Patchouli	30
Sandal wood	32
Geranium	33
Oil of cloves	34
Otto of roses	37
Bergamot	44
Neroli	47
Lavender	60
Lemon	65
Portugal	67
Thyme	68
Rosemary	74
Oil of laurel	80
Camomile flowers	87
Cassia	109
Spikenard	355
Aniseed	372

Thus the absorption by a tube full of dry air being 1, that of the odour of patchouli diffused in it is 30, that of lavender 60, that of rosemary 74, whilst that of aniseed amounts to 372. It would be idle to speculate on the quantities of matter concerned in these actions.

12. *Aqueous Vapour in relation to the Terrestrial Temperatures.*

We are now fully prepared for a result which, without such preparation, might appear incredible. Water is, to some extent, a volatile body, and our atmosphere, resting as it does upon the surface of the ocean, receives from it a continual supply of aqueous vapour. It would be an error to confound clouds or fog or any visible mist with the vapour of water, which is a perfectly impalpable gas, diffused, even on the clearest days, throughout the atmosphere. Compared with the great body of the air, the aqueous vapour it contains is of almost infinitesimal amount, $99\frac{1}{2}$ out of every 100 parts of the atmosphere being composed of oxygen and nitrogen. In the absence of experiment, we should never think of ascribing to this scant and varying constituent any important influence on terrestrial radiation; and yet its influence is far more potent than that of the great body of the air. To say that on a day of average humidity in England, the atmospheric vapour exerts 100 times the action of the air itself, would certainly be an understatement of the fact. Comparing a single molecule of aqueous vapour with an atom of either of the main constituents of our atmosphere, I am not prepared to say how many thousand times the action of the former exceeds that of the latter.

But it must be borne in mind that these large numbers depend, in part, on the extreme feebleness of the air; the power of aqueous vapour seems vast, because that of the air with which it is compared is infinitesimal. Absolutely considered, however, this substance, notwithstanding its small specific gravity,

exercises a very potent action. Probably from 10 to
15 per cent. of the heat radiated from the earth is
absorbed within 10 or 20 feet of the earth's surface.
This must evidently be of the utmost consequence to
the life of the world. Imagine the superficial molecules
of the earth agitated with the motion of heat, and
imparting it to the surrounding ether; this motion
would be carried rapidly away, and lost for ever to our
planet, if the waves of ether had nothing but the air
to contend with in their outward course. But the
aqueous vapour takes up the motion, and becomes
thereby heated, thus wrapping the earth like a warm
garment, and protecting its surface from the deadly
chill which it would otherwise sustain. Various philosophers have speculated on the influence of an atmospheric envelope. De Saussure, Fourier, M. Pouillet,
and Mr. Hopkins have, one and all, enriched scientific
literature with contributions on this subject, but the
considerations which these eminent men have applied
to atmospheric air, have, if my experiments be correct,
to be transferred to the aqueous vapour.

. The observations of meteorologists furnish important, though hitherto unconscious evidence of the
influence of this agent. Wherever the air is dry we
are liable to daily extremes of temperature. By day,
in such places, the sun's heat reaches the earth unimpeded, and renders the maximum high; by night, on
the other hand, the earth's heat escapes unhindered
into space, and renders the minimum low. Hence the
difference between the maximum and minimum is
greatest where the air is driest. In the plains of India,
on the heights of the Himalaya, in central Asia, in
Australia—wherever drought reigns, we have the heat
of day forcibly contrasted with the chill of night. In
the Sahara itself, when the sun's rays cease to impinge

on the burning soil, the temperature runs rapidly down to freezing, because there is no vapour overhead to check the calorific drain. And here another instance might be added to the numbers already known, in which nature tends as it were to check her own excess. By nocturnal refrigeration, the aqueous vapour of the air is condensed to water on the surface of the earth ; and, as only the superficial portions radiate, the act of condensation makes water the radiating body. Now experiment proves that to the rays emitted by water, aqueous vapour is especially opaque. Hence the very act of condensation, consequent on terrestrial cooling, becomes a safeguard to the earth, imparting to its radiation that particular character which renders it most liable to be prevented from escaping into space.

It might however be urged that, inasmuch as we derive all our heat from the sun, the selfsame covering which protects the earth from chill must also shut out the solar radiation. This is partially true, but only partially ; the sun's rays are different in quality from the earth's rays, and it does not at all follow that the substance which absorbs the one must necessarily absorb the other. Through a layer of water, for example, one tenth of an inch in thickness, the sun's rays are transmitted with comparative freedom ; but through a layer half this thickness, as Melloni has proved, no single ray from the warmed earth could pass. In like manner, the sun's rays pass with comparative freedom through the aqueous vapour of the air: the absorbing power of this substance being mainly exerted upon the invisible heat that endeavours to escape from the earth. In consequence of this differential action upon solar and terrestrial heat, the mean temperature of our planet is higher than is due to its distance from the sun.

13. *Liquids and their Vapours in relation to Radiant Heat.*

The deportment here assigned to atmospheric vapour has been established by direct experiments on air taken from the streets and parks of London, from the downs of Epsom, from the hills and sea-beach of the Isle of Wight, and also by experiments on air in the first instance dried, and afterwards rendered artificially humid by pure distilled water. It has also been established in the following way: Ten volatile liquids were taken at random and the power of these liquids, at a common thickness, to intercept the waves of heat, was carefully determined. The vapours of the liquids were next taken, in quantities proportional to the quantities of liquid, and the power of the vapours to intercept the waves of heat was also determined. Commencing with the substance which exerted the least absorptive power, and proceeding onwards to the most energetic, the following order of absorption was observed :—

Liquids	Vapours
Bisulphide of carbon.	Bisulphide of carbon.
Chloroform.	Chloroform.
Iodide of methyl.	Iodide of methyl.
Iodide of ethyl.	Iodide of ethyl.
Benzol.	Benzol.
Amylene.	Amylene.
Sulphuric ether.	Sulphuric ether.
Acetic ether.	Acetic ether.
Formic ether.	Formic ether.
Alcohol.	Alcohol.
Water.	

We here find the order of absorption in both cases to be the same. We have liberated the molecules from the bonds which trammel them more or less in a liquid condition; but this change in their state of aggregation does not change their relative powers of absorption. Nothing could more clearly prove that the act of absorption depends upon the individual molecule, which equally asserts its power in the liquid and the gaseous state. We may safely conclude from the above table that the position of a vapour is determined by that of its liquid. Now at the very foot of the list of liquids stands *water*, signalising itself above all others by its enormous power of absorption. And from this fact, even if no direct experiment on the vapour of water had ever been made, we should be entitled to rank that vapour as our most powerful absorber of radiant heat. Its attenuation, however, diminishes its action. I have proved that a shell of air two inches in thickness surrounding our planet, and saturated with the vapour of sulphuric ether, would intercept 35 per cent. of the earth's radiation. And though the quantity of aqueous vapour necessary to saturate air is much less than the amount of sulphuric ether vapour which it can sustain, it is still extremely probable that the estimate already made of the action of atmospheric vapour within 10 feet of the earth's surface, is under the mark ; and that we are indebted to this wonderful substance, to an extent not accurately determined, but certainly far beyond what has hitherto been imagined, for the temperature now existing at the surface of the globe.

14. *Reciprocity of Radiation and Absorption.*

Throughout the reflections which have hitherto occupied us, the image before the mind has been that of a radiant source sending forth calorific waves, which on passing among the molecules of a gas or vapour were intercepted by those molecules in various degrees. In all cases it was the transference of motion from the ether to the comparatively quiescent molecules of the gas or vapour that occupied our thoughts. We have now to change the form of our conception, and to figure these molecules not as absorbers but as radiators, not as the recipients but as the originators of wave-motion. That is to say, we must figure them vibrating, and generating in the surrounding ether undulations which speed through it with the velocity of light. Our object now is to enquire whether the act of chemical combination, which proves so potent as regards the phenomena of absorption, does not also manifest its power in the phenomena of radiation. For the examination of this question it is necessary, in the first place, to heat our gases and vapours to the same temperature, and then examine their power of discharging the motion thus imparted to them upon the ether in which they swing.

A heated copper ball was placed above a ring gas-burner possessing a great number of small apertures, the burner being connected by a tube with vessels containing the various gases to be examined. By gentle pressure the gases were forced through the orifices of the burner against the copper ball, where each of them, being heated, rose in an ascending column. A thermo-electric pile, entirely screened from the hot ball, was exposed to the radiation of the warm gas, while the deflection of a magnetic needle connected with the pile declared the energy of the radiation.

By this mode of experiment it was proved that the selfsame molecular arrangement which renders a gas a powerful absorber, renders it a powerful radiator—that the atom or molecule which is competent to intercept the calorific waves is, in the same degree, competent to send them forth. Thus, while the atoms of elementary gases proved themselves unable to emit any sensible amount of radiant heat, the molecules of compound gases were shown to be capable of powerfully disturbing the surrounding ether. By special modes of experiment the same was proved to hold good for the vapours of volatile liquids, the radiative power of every vapour being found proportional to its absorptive power.

The method of experiment here pursued, though not of the simplest character, is still easy to grasp. When air is permitted to rush into an exhausted tube, the temperature of the air is raised to a degree equivalent to the *vis viva* extinguished.[1] Such air is said to be dynamically heated, and, if pure, it shows itself incompetent to radiate, even when a rock-salt window is provided for the passage of its rays. But if instead of being empty the tube contain a small quantity of vapour, the warmed air communicates its heat by contact to the vapour, the molecules of which convert into the radiant form the heat imparted to them by the atoms of the air. By this process also, which I have called Dynamic Radiation, the reciprocity of radiation and absorption has been conclusively proved.[2]

In the excellent researches of Leslie, De la Provostaye and Desains, and Balfour Stewart, the same reciprocity, as regards solid bodies, has been variously illustrated; while the labours, theoretical and experimental, of Kirchhoff have given this subject a wonderful expansion, and enriched it by applications of the highest kind. To their results are now to be added the foregoing, whereby gases and vapours, which have been hitherto thought inaccessible to ex-

[1] See page 15 for a definition of *vis viva.*
[2] When heated air imparts its motion to another gas or vapour, the transference of heat is accompanied by a change of vibrating period. The Dynamic Radiation of vapours is rendered possible by this transmutation of vibrations.

periments with the thermo-electric pile, are proved by it to exhibit the indissoluble duality of radiation and absorption, the influence of chemical combination on both being exhibited in the most decisive and extraordinary way.

15. *Influence of Vibrating Period and Molecluan Form. Physical Analysis of the Human Breath.*

In the foregoing experiments with gases and vapours we have employed throughout invisible rays, and found some of these bodies so impervious to radiant heat, that in lengths of a few feet they intercept every ray as effectually as a layer of pitch. The substances, however, which show themselves thus opaque to radiant heat are perfectly transparent to light. Now the rays of light differ from those of invisible heat merely in point of period, the former failing to affect the retina because their periods of recurrence are too slow. Hence, in some way or other, the transparency of our gases and vapours depends upon the periods of the waves which impinge upon them. What is the nature of this dependence? The admirable researches of Kirchhoff help us to an answer. The atoms and molecules of every gas have certain definite rates of oscillation, and those waves of ether are most copiously absorbed whose periods of recurrence synchronise with those of the atomic groups amongst which they pass. Thus, when we find the invisible rays absorbed and the visible ones transmitted by a layer of gas, we conclude that the oscillating periods of the atoms constituting the gaseous molecules coincide with those of the invisible, and not with those of the visible spectrum.

It requires some discipline of the imagination to form a clear picture of this process. Such a picture is, however, possible, and ought to be obtained. When the waves of ether impinge upon molecules whose periods of vibration coincide with the recurrence of the undulations, the timed strokes of the waves augment the vibration of the molecules, as a heavy pendulum is set in motion by well-timed puffs of breath. Millions of millions of shocks are received every second from the calorific waves; and it is not difficult to see that as every wave arrives just in time to repeat the action of its predecessor, the molecules must finally be caused to swing through wider spaces than if the arrivals were not so timed. In fact, it is not difficult to see that an assemblage of molecules, operated upon by contending waves, might remain practically quiescent. This is actually the case when the waves of the visible spectrum pass through a transparent gas or vapour. There is here no sensible transference of motion from the ether to the molecules; in other words, there is no sensible absorption of heat.

One striking example of the influence of period may be here recorded. Carbonic acid gas is one of the feeblest absorbers of the radiant heat emitted by solid bodies. It is, for example, to a great extent transparent to the rays emitted by the heated copper plate already referred to. There are, however, certain rays, comparatively few in number, emitted by the copper, to which the carbonic acid is impervious; and could we obtain a source of heat emitting such rays only, we should find carbonic acid more opaque to the radiation from that source, than any other gas. Such a source is actually found in the flame of carbonic oxide, where hot carbonic acid constitutes the main radiating body. Of the rays emitted by our heated plate of copper, olefiant gas absorbs ten times the quantity absorbed by carbonic acid. Of the rays emitted by a carbonic oxide flame, carbonic acid absorbs twice as much as olefiant gas. This wonderful change in the power of the former, as an absorber, is simply due to the fact, that the periods of the hot and cold carbonic acid are identical, and that the waves from the flame freely transfer their motion to the molecules which synchronise with them. Thus it is that the tenth of an atmosphere of carbonic acid, enclosed in a tube four feet long, absorbs 60 per cent. of the radiation from a carbonic oxide flame, while one-thirtieth of an atmosphere absorbs 48 per cent. of the heat from the same source.

In fact, the presence of the minutest quantity of carbonic acid may be detected by its action on the rays from the carbonic oxide flame. Carrying, for example, the dried human breath into a tube four feet long, the absorption there effected by the carbonic acid of the breath amounts to 50 per cent. of the entire radiation. Radiant heat may indeed be employed as a means of determining practically the amount of carbonic acid expired from the lungs. My late assistant, Mr. Barrett, while under my direction, made this determination. The absorption produced by the breath freed from its moisture, but retaining its carbonic acid, was first determined. Carbonic acid, artificially prepared, was then mixed with dry air in such proportions that the action of the mixture upon the rays of heat was the same as that of the dried breath. The percentage of the former being known, immediately gave that of the latter. The same breath, analysed chemically by Dr. Frankland, and physically by Mr. Barrett, gave the following results :—

Percentage of Carbonic Acid in the Human Breath.

Chemical analysis				Physical analysis
4·66	.	.	.	4·56
5·33	.	.	.	5·22

It is thus proved that in the quantity of ethereal motion which it is competent to take up, we have a practical measure of the carbonic acid of the breath, and hence of the combustion going on in the human lungs.

Still this question of period, though of the utmost importance, is not competent to account for the whole of the observed facts. The ether, as far as we know, accepts vibrations of all periods with the same readiness. To it the oscillations of an atom of free oxygen are just as acceptable as those of the atoms in a molecule of olefiant gas ; that the vibrating oxygen then stands so far below the olefiant gas in radiant power must be referred not to period, but to some other peculiarity. The atomic group which constitutes the molecule of olefiant gas, produces many thousand times the disturbance caused by the oxygen, it may be because the group is able to lay a vastly more powerful hold upon the ether than the single atoms can. Another, and probably very potent cause of the difference may be, that the vibrations, being those of the constituent atoms of the molecule,[1] are generated in highly condensed ether, which acts like condensed air upon sound. But whatever may be the fate of these attempts to visualise the physics of the process, it will still remain true, that to account for the phenomena of radiation and absorption we must take into consideration the shape, size, and condition of the ether within the molecules, by which the external ether is disturbed.

[1] See ' Physical Considerations,' Art. iv. p. 102.

16. *Summary and Conclusion.*

Let us now cast a momentary glance over the ground that we have left behind. The general nature of light and heat was first briefly described : the compounding of matter from elementary atoms, and the influence of the act of combination on radiation and absorption, were considered and experimentally illustrated. Through the transparent elementary gases radiant heat was found to pass as through a vacuum, while many of the compound gases presented almost impassable obstacles to the calorific waves. This deportment of the simple gases directed our attention to other elementary bodies, the examination of which led to the discovery that the element iodine, dissolved in bisulphide of carbon, possesses the power of detaching, with extraordinary sharpness, the light of the spectrum from its heat, intercepting all luminous rays up to the extreme red, and permitting the calorific rays beyond the red to pass freely through it. This substance was then employed to filter the beams of the electric light, and to form foci of invisible rays so intense as to produce almost all the effects obtainable in an ordinary fire. Combustible bodies were burnt, and refractory ones were raised to a white heat, by the concentrated invisible rays. Thus, by exalting their refrangibility, the invisible rays of the electric light were rendered visible, and all the colours of the solar spectrum were extracted from utter darkness. The extreme richness of the electric light in invisible rays of low refrangibility was demonstrated, one-eighth only of its radiation consisting of luminous rays. The deadness of the optic nerve to those invisible rays was proved, and experiments were then added to show that the bright and the dark rays of a solid body, raised gradually to incandescence, are strengthened together; intense dark heat being an invariable accompaniment of intense white heat. A sun could not be formed, or a meteorite rendered luminous, on any other condition. The light-giving rays constituting only a small fraction of the total radiation, their unspeakable importance to us is due to the fact, that their periods are attuned to the special requirements of the eye.

Among the vapours of volatile liquids vast differences were also found to exist, as regards their powers of absorption. We followed various molecules from a state of liquid to a state of gas, and found, in both states of aggregation, the power of the individual molecules equally asserted. The position of a vapour as an absorber of radiant heat was shown to be determined by that of the liquid from which it is derived. Reversing our conceptions, and regarding the molecules of gases and vapours not as the recipients but as the originators of wave-motion; not as absorbers but as radiators; it was proved that the powers of absorption and radiation went hand in hand, the self-same chemical act which rendered a body competent to intercept the waves of ether, rendering it competent, in the same degree, to generate them. Perfumes were next subjected to examination, and, notwithstanding their extraordinary tenuity, they were found vastly superior, in point of absorptive power, to the body of the air in which they were diffused. We were led thus slowly up to the examination of the most widely diffused and most important of all vapours—the aqueous vapour of our atmosphere, and we found in it a potent absorber of the purely calorific rays. The power of this substance to influence climate, and its general influence on the temperature of the earth, were then briefly dwelt upon. A cobweb spread above a blossom is sufficient to protect it from nightly chill; and thus the aqueous vapour of our air, attenuated as it is, checks the drain of terrestrial heat, and saves the surface of our planet from the refrigeration which would assuredly accrue, were no such substance interposed between it and the voids of space. We considered the influence of vibrating period, and molecular form, on absorption and radiation, and finally deduced, from its action upon radiant heat, the exact amount of carbonic acid expired by the human lungs.

Thus, in brief outline, were placed before you some of the results of recent enquiries in the domain of Radiation, and my aim throughout has been to raise in your minds distinct physical images of the various processes involved in our researches. It is thought by some that natural science has a deadening influence on the imagination, and a doubt might fairly be raised as to the value of any study which would necessarily have this effect. But the experience of the last hour must, I think, have convinced you, that the study of natural science goes hand in hand with the culture of the imagination. Throughout the greater part of this discourse we have been sustained by this faculty. We have been picturing atoms, and molecules, and vibrations, and waves, which eye has never seen nor ear heard, and which can only be discerned by the exercise of imagination. This, in fact, is the faculty which enables us to transcend the boundaries of sense, and connect the phenomena of our visible world with those of an invisible one. Without imagination we never could have risen to the conceptions which have occupied us here to-day; and in proportion to your power of exercising this faculty aright, and of associating definite mental images with the terms employed, will be the pleasure and the profit which you will derive from this lecture.

The outward facts of nature are insufficient to satisfy the mind. We cannot be content with knowing that the light and heat of the sun illuminate and warm the world. We are led irresistibly to enquire, ' What is light, and what is heat?' and this question leads us at once out of the region of sense into that of imagination.[1]

Thus pondering, and questioning, and striving to supplement . that which is felt and seen, but which is incomplete, by something unfelt and unseen which is necessary to its completeness, men of genius have in part discerned, not only the nature of light and heat, but also, through them, the general relationship of natural phenomena. The working power of Nature consists of actual or potential motion, of which all its phenomena are but special forms. This motion manifests itself in tangible and in intangible matter, being incessantly transferred from the one to the other, and incessantly transformed by the change. It is as real in the waves of the ether as in the waves of the sea; the latter—derived as they are from winds, which in their turn are derived from the sun—are, indeed, nothing more than the heaped-up motion of the ether waves. It is the calorific waves emitted by the sun which heat our air, produce our winds, and hence agitate our ocean. And whether they break in foam upon the shore, or rub silently against the ocean's bed, or subside by the mutual friction of their own parts, the sea waves, which cannot subside without producing heat, finally resolve themselves into waves of ether, thus regenerating the motion from which their temporary existence was derived. This connection is typical. Nature is not an aggregate of independent parts, but an organic whole. If you open a piano and sing into

[1] This line of thought was pursued further five years subsequently. See ' Scientific Use of the Imagination ' in Vol. II.

it, a certain string will respond. Change the pitch of your voice; the first string ceases to vibrate, but another replies. Change again the pitch; the first two strings are silent, while another resounds. Thus is sentient man acted on by Nature, the optic, the auditory, and other nerves of the human body being so many strings differently tuned, and responsive to different forms of the universal power.

III.

ON RADIANT HEAT IN RELATION TO THE COLOUR AND CHEMICAL CONSTITUTION OF BODIES.[1]

ONE of the most important functions of physical science, considered as a discipline of the mind, is to enable us by means of the sensible processes of Nature to apprehend the insensible. The sensible processes give direction to the line of thought; but this once given, the length of the line is not limited by the boundaries of the senses. Indeed, the domain of the senses, in Nature, is almost infinitely small in comparison with the vast region accessible to thought which lies beyond them. From a few observations of a comet, when it comes within the range of his telescope, an astronomer can calculate its path in regions which no telescope can reach : and in like manner, by means of data furnished in the narrow world of the senses, we make ourselves at home in other and wider worlds, which are traversed by the intellect alone.

From the earliest ages the questions, 'What is light?' and 'What is heat?' have occurred to the minds of men; but these questions never would have been answered had they not been preceded by the question, 'What is sound?' Amid the grosser phenomena of acoustics the mind was first disciplined, conceptions

[1] A discourse delivered in the Royal Institution of Great Britain, Jan. 19, 1866.

being thus obtained from direct observation, which were afterwards applied to phenomena of a character far too subtle to be observed directly. Sound we know to be due to vibratory motion. A vibrating tuning-fork, for example, moulds the air around it into undulations or waves, which speed away on all sides with a certain measured velocity, impinge upon the drum of the ear, shake the auditory nerve, and awake in the brain the sensation of sound. When sufficiently near a sounding body we can feel the vibrations of the air. A deaf man, for example, plunging his hand into a bell when it is sounded, feels through the common nerves of his body those tremors which, when imparted to the nerves of healthy ears, are translated into sound. There are various ways of rendering those sonorous vibrations not only tangible but visible; and it was not until numberless experiments of this kind had been executed, that the scientific investigator abandoned himself wholly, and without a shadow of misgiving, to the conviction that what is sound within us is, outside of us, a motion of the air.

But once having established this fact—once having proved beyond all doubt that the sensation of sound is produced by an agitation of the auditory nerve—the thought soon suggested itself that light might be due to an agitation of the optic nerve. This was a great step in advance of that ancient notion which regarded light as something emitted by the eye, and not as anything imparted to it. But if light be produced by an agitation of the retina, what is it that produces the agitation? Newton, you know, supposed minute particles to be shot through the humours of the eye against the retina, which he supposed to hang like a target at the back of the eye. The impact of these particles against the target, Newton believed to be the cause of light. But Newton's notion has not held its ground, being entirely driven from the field by the more wonderful and far more philosophical notion that light, like sound, is a product of wave-motion.

The domain in which this motion of light is carried on lies entirely beyond the reach of our senses. The waves of light require a medium for their formation and propagation ; but we cannot see, or feel, or taste, or smell this medium. How, then, has its existence been established ? By showing, that by the assumption of this wonderful intangible *ether*, all the phenomena of optics are accounted for, with a fulness, and clearness, and conclusiveness, which leave no desire of the intellect unsatisfied. When the law of gravitation first suggested itself to the mind of Newton, what did he do ? He set himself to examine whether it accounted for all the facts. He determined the courses of the planets ; he calculated the rapidity of the moon's fall towards the earth ; he considered the precession of the equinoxes, the ebb and flow of the tides, and found all explained by the law of gravitation. He therefore regarded this law as established, and the verdict of science subsequently confirmed his conclusion. On similar, and, if possible, on stronger grounds, we found our belief in the existence of the universal ether. It explains facts far more various and complicated than those on which Newton based his law. If a single phenomenon could be pointed out which the ether is proved incompetent to explain, we should have to give it up ; but no such phenomenon has ever been pointed out. It is, therefore, at least as certain that space is filled with a medium, by means of which suns and stars diffuse their radiant power, as that it is traversed by that force which holds in its grasp, not only our planetary system, but the immeasurable heavens themselves.

There is no more wonderful instance than this of the production of a line of thought, from the world of the senses into the region of pure imagination. I mean by imagination here, not that play of fancy which can give to airy nothings a local habitation and a name, but that power which enables the mind to conceive realities which lie beyond the range of the senses—to present to itself distinct images of processes which, though mighty in the aggregate beyond all conception, are so minute individually as to elude all observation. It is the waves of air excited by a tuning-fork which render its vibrations audible. It is the waves of ether sent forth from those lamps overhead which render them luminous to us; but so minute are these waves, that it would take from 30,000 to 60,000 of them placed end to end to cover a single inch. Their number, however, compensates for their minuteness. Trillions of them have entered your eyes, and hit the retina at the backs of your eyes, in the time consumed in the utterance of the shortest sentence of this discourse. This is the steadfast result of modern research; but we never could have reached it without previous discipline. We never could have measured the waves of light, nor even imagined them to exist, had we not previously exercised ourselves among the waves of sound. Sound and light are now mutually helpful, the conceptions of each being expanded, strengthened, and defined by the conceptions of the other.

The ether which conveys the pulses of light and heat not only fills celestial space, swathing suns, and planets, and moons, but it also encircles the atoms of which these bodies are composed. It is the motion of these atoms, and not that of any sensible parts of bodies, that the ether conveys. This motion is the objective cause of what, in our sensations, are light and heat. An atom, then, sending its pulses through the ether, resembles a tuning-fork sending its pulses through the air. Let us look for a moment at this thrilling medium, and briefly consider its relation to the bodies whose vibrations it conveys. Different bodies, when heated to the same temperature, possess very different powers of agitating the ether: some are good radiators, others are bad radiators ; which means that some are so constituted as to communicate their atomic motion freely to the ether, producing therein powerful undulations ; while the atoms of others are unable thus to communicate their motions, but glide through the medium without materially disturbing its repose. Recent experiments have proved that elementary bodies, except under certain anomalous conditions, belong to the class of bad radiators. An atom, vibrating in the ether, resembles a naked tuning-fork vibrating in the air. The amount of motion communicated to the air by the thin prongs is too small to evoke at any distance the sensation of sound. But if we permit the atoms to combine chemically and form molecules, the result, in many cases, is an enormous change in the power of radiation. The amount of ethereal disturbance, produced by the combined atoms of a body, may be many thousand times that produced by the same atoms when uncombined.

The pitch of a musical note depends upon the rapidity of its vibrations, or, in other words, on the length of its waves. Now, the pitch of a note answers to the colour of light. Taking a slice of white light from the sun, or from an electric lamp, and causing the light to pass through an arrangement of prisms, it is decomposed. We have the effect obtained by Newton,

who first unrolled the solar beam into the splendours of
the solar spectrum.   At one end of this spectrum we
have red light, at the other, violet ; and between those
extremes lie the other prismatic colours.   As we advance
along the spectrum from the red to the violet, the
pitch of the light—if I may use the expression—
heightens, the sensation of violet being produced by
a more rapid succession of impulses than that which
produces the impression of red.   The vibrations of the
violet are about twice as rapid as those of the red ; in
other words, the range of the visible spectrum is about
an octave.

There is no solution of continuity in this spectrum ;
one colour changes into another by insensible gradations.
It is as if an infinite number of tuning-forks, of gradually augmenting pitch, were vibrating at the same time.
But turning to another spectrum—that, namely, obtained from the incandescent vapour of silver—you
observe that it consists of two narrow and intensely
luminous green bands.   Here it is as if two forks only,
of slightly different pitch, were vibrating.   The length
of the waves which produce this first band is such that
47,460 of them, placed end to end, would fill an inch.
The waves which produce the second band are a little
shorter; it would take of these 47,920 to fill an inch.
In the case of the first band, the number of impulses
imparted, in one second, to every eye which sees it, is
577 millions of millions ; while the number of impulses
imparted, in the same time, by the second band is 600
millions of millions.   We may project upon a white
screen the beautiful stream of green light from which
these bands were derived.   This luminous stream is the
incandescent vapour of silver.   The rates of vibration
of the atoms of that vapour are as rigidly fixed as those
of two tuning-forks ; and to whatever height the tem-

perature of the vapour may be raised, the rapidity of its vibrations, and consequently its colour, which wholly depends upon that rapidity, remain unchanged.

The vapour of water, as well as the vapour of silver, has its definite periods of vibration, and these are such as to disqualify the vapour, when acting freely as such, from being raised to a white heat. The oxyhydrogen flame, for example, consists of hot aqueous vapour. It is scarcely visible in the air of this room, and it would be still less visible if we could burn the gas in a clean atmosphere. But the atmosphere, even at the summit of Mont Blanc, is dirty; in London it is more than dirty; and the burning dirt gives to this flame the greater portion of its present light. But the heat of the flame is enormous. Cast iron fuses at a temperature of 2,000° Fahr.; while the temperature of the oxyhydrogen flame is 6,000° Fahr. A piece of platinum is heated to vivid redness, at a distance of two inches beyond the visible termination of the flame. The vapour which produces incandescence is here absolutely dark. In the flame itself the platinum is raised to dazzling whiteness, and is even pierced by the flame. When this flame impinges on a piece of lime, we have the dazzling Drummond light. But the light is here due to the fact that when it impinges upon the solid body, the vibrations excited in that body by the flame are of periods different from its own.

Thus far we have fixed our attention on atoms and molecules in a state of vibration, and surrounded by a medium which accepts their vibrations, and transmits them through space. But suppose the waves generated by one system of molecules to impinge upon another system, how will the waves be affected? Will they be stopped, or will they be permitted to pass? Will they transfer their motion to the molecules on which they impinge, or will they glide round the molecules, through the intermolecular spaces, and thus escape?

The answer to this question depends upon a condition which may be beautifully exemplified by an experiment on sound. These two tuning-forks are tuned absolutely alike. They vibrate with the same rapidity, and, mounted thus upon their resonant cases, you hear them loudly sounding the same musical note. Stopping one of the forks, I throw the other into strong vibration, and bring that other near the silent fork, but not into contact with it. Allowing them to continue in this position for four or five seconds, and then stopping the vibrating fork, the sound does not cease. The second fork has taken up the vibrations of its neighbour, and is now sounding in its turn. Dismounting one of the forks, and permitting the other to remain upon its stand, I throw the dismounted fork into strong vibration. You cannot hear it sound. Detached from its case, the amount of motion which it can communicate to the air is too small to be sensible at any distance. When the dismounted fork is brought close to the mounted one, but not into actual contact with it, out of the silence rises a mellow sound. Whence comes it? From the vibrations which have been transferred from the dismounted fork to the mounted one.

That the motion should thus transfer itself through the air it is necessary that the two forks should be in perfect unison. If a morsel of wax not larger than a pea be placed on one of the forks, it is rendered thereby powerless to affect, or to be affected by, the other. It is easy to understand this experiment. The pulses of the one fork can affect the other, because they are *perfectly timed*. A single pulse causes the prong of the silent fork to vibrate through an infinitesimal space. But just as it has completed this small vibration, another pulse is ready to strike it. Thus, the impulses add themselves together. In the five seconds during which the forks were held near each other, the vibrating fork sent 1,280 waves against its neighbour and those 1,280 shocks, all delivered at the proper moment, all, as I have said, perfectly timed, have given such strength to the vibrations of the mounted fork as to render them audible to all.

Another curious illustration of the influence of synchronism on musical vibrations, is this : Three small gas-flames are inserted into three glass tubes of different lengths. Each of these flames can be caused to emit a musical note, the pitch of which is determined by the length of the tube surrounding the flame. The shorter the tube the higher is the pitch. The flames are now silent within their respective tubes, but each of them can be caused to respond to a proper note sounded anywhere in this room. With an instrument called a syren, a powerful musical note, of gradually increasing pitch, can be produced. Beginning with a low note, and ascending gradually to a higher one, we finally attain the pitch of the flame in the longest tube. The moment it is reached, the flame bursts into song. The other flames are still silent within their tubes. But by urging the instrument on to higher notes, the second flame is started, and the third alone remains. A still higher note starts it also. Thus, as the sound of the syren rises gradually in pitch, it awakens every flame in passing, by striking it with a series of waves whose periods of recurrence are similar to its own.

Now the wave-motion from the syren is in part taken up by the flame which synchronises with the waves ; and were these waves to impinge upon a multitude of flames, instead of upon one flame only, the transference might be so great as to absorb the whole of the original wave motion. Let us apply these facts to radiant heat. This blue flame is the flame of carbonic oxide; this transparent gas is carbonic acid gas. In the blue flame we have carbonic acid intensely heated, or, in other words, in a state of intense vibration. It thus resembles the sounding fork, while this cold carbonic acid resembles the silent one. What is the consequence? Through the synchronism of the hot and cold gas, the waves emitted by the former are intercepted by the latter, the transmission of the radiant heat being thus prevented. The cold gas is intensely opaque to the radiation from this particular flame, though highly transparent to heat of every other kind. We are here manifestly dealing with that great principle which lies at the basis of spectrum analysis, and which has enabled scientific men to determine the substances of which the sun, the stars, and even the nebulæ are composed; the principle, namely, that a body which is competent to emit any ray, whether of heat or light, is competent in the same degree to absorb that ray. The absorption depends on the synchronism existing between the vibrations of the atoms from which the rays, or more correctly the waves, issue, and those of the atoms on which they impinge.

To its almost total incompetence to emit white light, aqueous vapour adds a similar incompetence to absorb white light. It cannot, for example, absorb the luminous rays of the sun, though it can absorb the non-luminous rays of the earth. This incompetence of the vapour to absorb luminous rays is shared by water and ice—in fact, by all really transparent substances. Their transparency is due to their inability to absorb luminous rays. The molecules of such substances are in dissonance with the luminous waves; and hence such waves pass through transparent bodies without disturbing the molecular rest. A purely luminous beam, however intense may be its heat, is sensibly incompetent to melt ice. We can, for example, converge a powerful luminous beam upon a surface covered with hoar frost, without melting a single spicula of the crystals. How then, it may be asked, are the snows of the Alps swept away by the sunshine of summer? I answer, they are not swept away by sunshine at all, but by rays which have no sunshine whatever in them. The luminous rays of the sun fall upon the snow-fields and are flashed in echoes from crystal to crystal, but they find next to no lodgment within the crystals. They are hardly at all absorbed, and hence they cannot produce fusion. But a body of powerful dark rays is emitted by the sun ; and it is these that cause the glaciers to shrink and the snows to disappear ; it is they that fill the banks of the Arve and Arveyron, and liberate from their frozen captivity the Rhone and the Rhine.

Placing a concave silvered mirror behind the electric light its rays are converged to a focus of dazzling brilliancy. Placing in the path of the rays, between the light and the focus, a vessel of water, and introducing at the focus a piece of ice, the ice is not melted by the concentrated beam. Matches, at the same place, are ignited, and wood is set on fire. The powerful heat, then, of this luminous beam is incompetent to melt the ice. On withdrawing the cell of water, the ice immediately liquefies, and the water trickles from it in drops. Reintroducing the cell of water, the fusion is arrested, and the drops cease to fall. The transparent water of the cell exerts no sensible absorption on the luminous rays, still it withdraws something from the beam, which, when permitted to act, is competent to melt the ice. This something is the dark radiation of the electric light. Again, I place a slab of pure ice in front of the electric lamp ; send a luminous beam first through our cell of water and then through the ice. By means of a lens an image of the slab is cast upon a white screen. The beam, sifted by the water, has little power upon the ice. But observe what occurs when the water is removed; we have here a star and there a star, each star resembling a flower of six petals, and growing visibly larger before our eyes. As the leaves enlarge, their edges become serrated, but there is no deviation from the six-rayed type. We have here, in fact, the crystallisation of the ice reversed by the invisible rays of the electric beam. They take the molecules down in this wonderful way, and reveal to us the exquisite atomic structure of the substance with which Nature every winter roofs our ponds and lakes.

Numberless effects, apparently anomalous, might be adduced in illustration of the action of these lightless rays. These two powders, for example, are both white, and undistinguishable from each other by the eye. The luminous rays of the sun are unabsorbed by both—from such rays these powders acquire no heat; still one of them, sugar, is heated so highly by the concentrated beam of the electric lamp, that it first smokes and then violently inflames, while the other substance, salt, is barely warmed at the focus. Placing two perfectly transparent liquids in test-tubes at the focus, one of them boils in a couple of seconds, while the other, in a similar position, is hardly warmed. The boiling-point of the first liquid is 78° C., which is speedily reached; that of the second liquid is only 48° C., which is never reached at all. These anomalies are entirely due to the unseen element which mingles with the luminous rays of the electric beam, and indeed constitutes 90 per cent. of its calorific power.

A substance, as many of you know, has been discovered, by which these dark rays may be detached from the total emission of the electric lamp. This ray-filter is a liquid, black as pitch to the luminous, but bright as a diamond to the non-luminous, radiation. It mercilessly cuts off the former, but allows the latter free transmission. When these invisible rays are brought to a focus, at a distance of several feet from the electric lamp, the dark rays form an invisible image of their source. By proper means, this image may be transformed into a visible one of dazzling brightness. It might, moreover, be shown, if time permitted, how, out of those perfectly dark rays, could be extracted, by a process of transmutation, all the colours of the solar spectrum. It might also be proved that those rays, powerful as they are, and sufficient to fuse many metals, can be permitted to enter the eye, and to break upon the retina, without producing the least luminous impression.

The dark rays being thus collected, you see nothing at their place of convergence. With a proper thermometer it could be proved that even the air at the focus is just as cold as the surrounding air. And mark the conclusion to which this leads. It proves the ether at the focus to be practically detached from the air,—that the most violent ethereal motion may there exist, without the least aërial motion. But, though you see it not, there is sufficient heat at that focus to set London on fire. The heat there is competent to raise iron to a temperature at which it throws off brilliant scintillations. It can heat platinum to whiteness, and almost fuse that refractory metal. It actually can fuse gold, silver, copper, and aluminium. The moment, moreover, that wood is placed at the focus it bursts into a blaze.

It has been already affirmed that, whether as regards radiation or absorption, the elementary atoms possess but little power. This might be illustrated by a long array of facts; and one of the most singular of these is furnished by the deportment of that extremely combustible substance, phosphorus, when placed at the dark focus. It is impossible to ignite there a fragment of amorphous phosphorus. But ordinary phosphorus is a far quicker combustible, and its deportment towards radiant heat is still more impressive. It may be exposed to the intense radiation of an ordinary fire without bursting into flame. It may also be exposed for twenty or thirty seconds at an obscure focus, of sufficient power to raise platinum to a red heat, without ignition. Notwithstanding the energy of the ethereal waves here concentrated, notwithstanding the extremely inflammable character of the elementary body exposed to their action, the atoms of that body refuse to partake of the motion of the powerful waves of low refrangibility, and consequently cannot be affected by their heat.

The knowledge we now possess will enable us to analyse with profit a practical question. White dresses are worn in summer, because they are found to be cooler than dark ones. The celebrated Benjamin Franklin placed bits of cloth of various colours upon snow, exposed them to direct sunshine, and found that they sank to different depths in the snow. The black cloth sank deepest, the white did not sink at all. Franklin inferred from this experiment that black bodies are the best absorbers, and white ones the worst absorbers, of radiant heat. Let us test the generality of this conclusion. One of these two cards is coated with a very dark powder, and the other with a perfectly white one. I place the powdered surfaces before a fire, and leave them there until they have acquired as high a temperature as they can attain in this position. Which of the cards is then most highly heated? It

7

requires no thermometer to answer this question. Simply pressing the back of the card, on which the white powder is strewn, against the cheek or forehead, it is found intolerably hot. Placing the dark card in the same position, it is found cool. The white powder has absorbed far more heat than the dark one. This simple result abolishes a hundred conclusions which have been hastily drawn from the experiment of Franklin. Again, here are suspended two delicate mercurial thermometers at the same distance from a gas-flame. The bulb of one of them is covered by a dark substance, the bulb of the other by a white one. Both bulbs have received the radiation from the flame, but the white bulb has absorbed most, and its mercury stands much higher than that of the other thermometer. This experiment might be varied in a hundred ways : it proves that from the darkness of a body you can draw no certain conclusion regarding its power of absorption.

The reason of this simply is, that colour gives us intelligence of only one portion, and that the smallest one, of the rays impinging on the coloured body. Were the rays all luminous, we might with certainty infer from the colour of a body its power of absorption ; but the great mass of the radiation from our fire, our gas-flame, and even from the sun itself, consists of invisible calorific rays, regarding which colour teaches us nothing. A body may be highly transparent to the one class of rays, and highly opaque to the other. Thus the white powder, which has shown itself so powerful an absorber, has been specially selected on account of its extreme perviousness to the visible rays, and its extreme imperviousness to the invisible ones ; while the dark powder was chosen on account of its extreme transparency to the invisible, and its extreme opacity to the visible, rays. In the case of the radiation from our fire, about 98 per cent. of the whole emission con sists of invisible rays; the body, therefore, which was most opaque to these triumphed as an absorber, though that body was a white one.

And here it is worth while to consider the manner in which we obtain from natural facts what may be called their intellectual value. Throughout the processes of Nature we have interdependence and harmony; and the main value of physics, considered as a mental discipline, consists in the tracing out of this interdependence, and the demonstration of this harmony. The outward and visible phenomena are the counters of the intellect; and our science would not be worthy of its name and fame if it halted at facts, however practically useful, and neglected the laws which accompany and rule the phenomena. Let us endeavour, then, to extract from the experiment of Franklin all that it can yield, calling to our aid the knowledge which our predecessors have already stored. Let us imagine two pieces of cloth of the same texture, the one black and the other white, placed upon sunned snow. Fixing our attention on the white piece, let us enquire whether there is any reason to expect that it will sink in the snow at all. There is knowledge at hand which enables us to reply at once in the negative. There is, on the contrary, reason to expect that, after a sufficient exposure, the bit of cloth will be found on an eminence instead of in a hollow; that instead of a depression, we shall have a relative elevation of the bit of cloth. For, as regards the luminous rays of the sun, the cloth and the snow are alike powerless; the one cannot be warmed, nor the other melted, by such rays. The cloth is white and the snow is white, because their confusedly mingled fibres and particles are incompetent to absorb the luminous rays. Whether, then, the cloth will sink or not depends entirely upon the dark rays of the sun. Now the substance which absorbs these dark rays with the greatest avidity is ice,—or snow, which is merely ice in powder. Hence, a less amount of heat will be lodged in the cloth than in the surrounding snow. The cloth must therefore act as a shield to the snow on which it rests; and, in consequence of the more rapid fusion of the exposed snow, its shield must, in due time, be left behind, perched upon an eminence like a glacier-table.

But though the snow transcends the cloth, both as a radiator and absorber, it does not much transcend it. Cloth is very powerful in both these respects. Let us now turn our attention to the piece of black cloth, the texture and fabric of which I assume to be the same as that of the white. For our object being to compare the effects of colour, we must, in order to study this effect in its purity, preserve all the other conditions constant. Let us then suppose the black cloth to be obtained from the dyeing of the white. The cloth itself, without reference to the dye, is nearly as good an absorber of heat as the snow around it. But to the absorption of the dark solar rays by the undyed cloth, is now added the absorption of the whole of the luminous rays, and this great additional influx of heat is far more than sufficient to turn the balance in favour of the black cloth. The sum of its actions on the dark and luminous rays, exceeds the action of the snow on the dark rays alone. Hence the cloth will sink in the snow, and this is the complete analysis of Franklin's experiment.

Throughout this discourse the main stress has been laid on chemical constitution, as influencing most powerfully the phenomena of radiation and absorption.

With regard to gases and vapours, and to the liquids from which these vapours are derived, it has been proved by the most varied and conclusive experiments that the acts of radiation and absorption are *molecular* —that they depend upon chemical, and not upon mechanical, condition. In attempting to extend this principle to solids I was met by a multitude of facts, obtained by celebrated experimenters, which seemed flatly to forbid such an extension. Melloni, for example, had found the same radiant and absorbent power for chalk and lamp-black. MM. Masson and Courtépée had performed a most elaborate series of experiments on chemical precipitates of various kinds, and found that they one and all manifested the same power of radiation. They concluded from their researches, that when bodies are reduced to an extremely fine state of division, the influence of this state is so powerful as entirely to mask and override whatever influence may be due to chemical constitution.

But it appears to me that through the whole of these researches an oversight has run, the mere mention of which will show what caution is essential in the operations of experimental philosophy; while an experiment or two will make clear wherein the oversight consists. Filling a brightly polished metal cube with boiling water, I determine the quantity of heat emitted by two of the bright surfaces. As a radiator of heat one of them far transcends the other. Both surfaces appear to be metallic ; what, then, is the cause of the observed difference in their radiative power ? Simply this: one of the surfaces is coated with transparent gum, through which, of course, is seen the metallic lustre behind ; and this varnish, though so perfectly transparent to luminous rays, is as opaque as pitch, or lamp-black, to non-luminous ones. It is a powerful emitter of dark rays ; it is also a powerful absorber. While, therefore, at the present moment, it is copiously pouring forth radiant heat itself, it does not allow a single ray from the metal behind to pass through it. The varnish then, and not the metal, is the real radiator.

Now Melloni, and Masson, and Courtépée experimented thus: they mixed their powders and precipitates with gum-water, and laid them, by means of a brush, upon the surfaces of a cube like this. True, they saw their red powders red, their white ones white, and their black ones black, but they saw these colours *through the coat of varnish which surrounded every particle.* When, therefore, it was concluded that colour had no influence on radiation, no chance had been given to it of asserting its influence; when it was found that all chemical precipitates radiated alike, it was the radiation from a varnish, common to them all, which showed the observed constancy. Hundreds, perhaps thousands, of experiments on radiant heat have been performed in this way, by various enquirers, but the work will, I fear, have to be done over again. I am not, indeed, acquainted with an instance in which an oversight of so trivial a character has been committed by so many able men in succession, vitiating so large an amount of otherwise excellent work.

Basing our reasonings thus on demonstrated facts, we arrive at the extremely probable conclusion that the envelope of the particles, and not the particles themselves, was the real radiator in the experiments just referred to. To reason thus, and deduce their more or less probable consequences from experimental facts, is an incessant exercise of the student of physical science. But having thus followed, for a time, the light of reason alone through a series of phenomena, and emerged from them · with a purely intellectual conclusion, our duty is to bring that conclusion to an experimental test.   In this way we fortify our science.

For the purpose of testing our conclusion regarding the influence of the gum, I take two powders presenting the same physical appearance; one of them is a compound of mercury, and the other a compound of lead. On two surfaces of a cube are spread these bright red powders, without varnish of any kind.   Filling the cube with boiling water, and determining the radiation from the two surfaces, one of them is found to emit thirty-nine units of heat, while the other emits seventy-four.   This, surely, is a great difference.   Here, however, is a second cube, having two of its surfaces coated with the same powders, the only difference being that the powders are laid on by means of a transparent gum. Both surfaces are now absolutely alike in radiative power.   Both of them emit somewhat more than was emitted by either of the unvarnished powders, simply because the gum employed is a better radiator than either of them.   Excluding all varnish, and comparing white with white, vast differences are found ; comparing black with black, they are also different; and when black and white are compared, in some cases the black radiates far more than the white, while in other cases the white radiates far more than the black.   Determining, moreover, the absorptive power of those powders, it is found to go hand-in-hand with their radiative power.   The good radiator is a good absorber, and the bad radiator is a bad absorber.   From all this it is evident that as regards the radiation and absorption of non-luminous heat, colour teaches us nothing; and that even as regards the radiation of the sun, consisting as it does mainly of non-luminous rays, conclusions as to the influence of colour may be altogether delusive.   This is the strict scientific upshot of

our researches. But it is not the less true that in the
case of wearing apparel—and this for reasons which I
have given in analysing the experiment of Franklin—
black dresses are more potent than white ones as absorbers of solar heat.

Thus, in brief outline, have been brought before
you a few of the results of recent enquiry. If you ask
me what is the use of them, I can hardly answer you,
unless you define the term use. If you meant to ask
whether those dark rays which clear away the Alpine
snows, will ever be applied to the roasting of turkeys,
or the driving of steam-engines—while affirming their
power to do both, I would frankly confess that they
are not at present capable of competing profitably with
coal in these particulars. Still they may have great
uses unknown to me; and when our coal-fields are
exhausted, it is possible that a more ethereal race
than we are may cook their victuals, and perform their
work, in this transcendental way. But is it necessary
that the student of science should have his labours
tested by their possible practical applications? What
is the practical value of Homer's Iliad? You smile,
and possibly think that Homer's Iliad is good as a
means of culture. There's the rub. The people who
demand of science practical uses, forget, or do not
know, that it also is great as a means of culture—that
the knowledge of this wonderful universe is a thing
profitable in itself, and requiring no practical application to justify its pursuit.

But while the student of Nature distinctly refuses
to have his labours judged by their practical issues,
unless the term practical be made to include mental
as well as material good, he knows full well that the
greatest practical triumphs have been episodes in the
search after pure natural truth. The electric telegraph

is the standing wonder of this age, and the men whose
scientific knowledge, and mechanical skill, have made
the telegraph what it is, are deserving of all honour.
In fact, they have had their reward, both in reputation
and in those more substantial benefits which the direct
service of the public always carries in its train.   But
who, I would ask, put the soul into this telegraphic
body?   Who snatched from heaven the fire that flashes
along the line?   This, I am bound to say, was done
by two men, the one a dweller in Italy,[1] the other a
dweller in England,[2] who never in their enquiries
consciously set a practical object before them—whose
only stimulus was the fascination which draws the
climber to a never-trodden peak, and would have made
Cæsar quit his victories for the sources of the Nile.
That the knowledge brought to us by those prophets,
priests, and kings of science is what the world calls
' useful knowledge,' the triumphant application of their
discoveries proves.   But science has another function
to fulfil, in the storing and the training of the human
mind ; and I would base my appeal to you on the
specimen which has this evening been brought before
you, whether any system of education at the present
day can be deemed even approximately complete, in
which the knowledge of Nature is neglected or ignored.

[1] Volta.   [2] Faraday.

IV.

NEW CHEMICAL REACTIONS PRODUCED BY LIGHT.

1868–69.

MEASURED by their power, not to excite vision, but to produce heat—in other words, measured by their absolute energy—the ultra-red waves of the sun and of the electric light, as shown in the preceding articles, far transcend the visible. In the domain of chemistry, however, there are numerous cases in which the more powerful waves are ineffectual, while the more minute waves, through what may be called their timeliness of application, are able to produce great effects. A series of these, of a novel and beautiful character, discovered in 1868, and further illustrated in subsequent years, may be exhibited by subjecting the vapours of volatile liquids to the action of concentrated sunlight, or to the concentrated beam of the electric light. Their investigation led up to the discourse on ' Dust and Disease ' which follows in this volume; and for this reason some account of them is introduced here.

A glass tube 3 feet long and 3 inches wide, which had been frequently employed in my researches on radiant heat, was supported horizontally on two stands. At one end of the tube was placed an electric lamp, the height and position of both being so arranged, that the axis of the tube, and that of the beam issuing from the lamp, were coincident. In the first experiments the two ends of the tube were closed by plates of rock-salt, and subsequently by plates of glass. For the sake of distinction, I call this tube *the experimental tube.* It was connected with an air-pump, and also with a series of drying and other tubes used for the purification of the air.

A number of test-tubes, like F, fig. 2 (I have used at least fifty of them), were converted into Woulf's flasks. Each of them was stopped by a cork, through which passed two glass tubes: one of these tubes (a) ended immediately below the cork, while the other (b) descended to the bottom of the flask, being drawn out at its lower end to an orifice about 0·03 of an inch in diameter. It was found necessary to coat the cork carefully with cement. In the later experiments corks of vulcanised india-rubber were invariably employed.

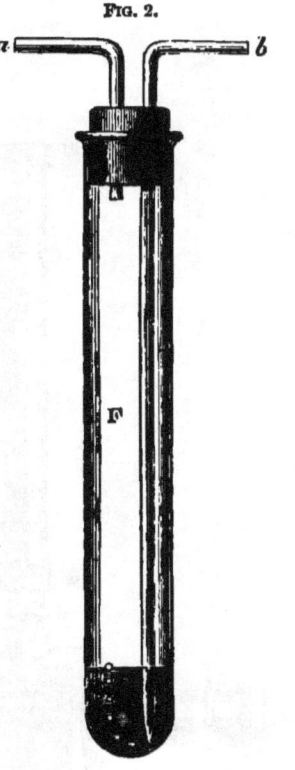

Fig. 2.

The little flask, thus formed, being partially filled with the liquid whose vapour was to be examined, was introduced into the path of the purified current of air. The experimental tube being exhausted, and the cock which cut off the supply of purified air being cautiously turned on, the air entered the flask through the tube b, and escaped by the small orifice at the lower end of b into the liquid. Through this it bubbled, loading

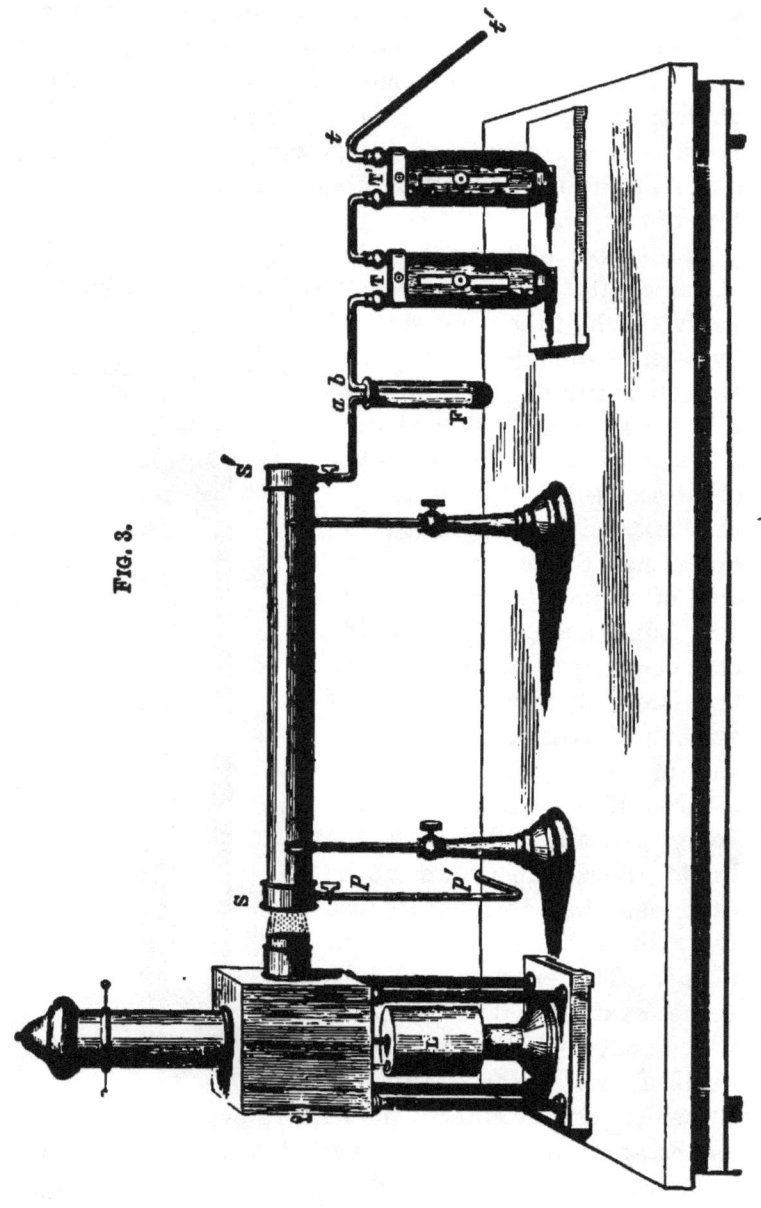

FIG. 3.

itself with vapour, after which the mixed air and vapour, passing from the flask by the tube *a*, entered the experimental tube, where they were subjected to the action of light.

The whole arrangement is shown in fig. 3, where L represents the electric lamp, s s′ the experimental tube, *p p′* the pipe leading to the air-pump, and F the test-tube containing the volatile liquid. The tube *t t′* is plugged with cotton-wool intended to intercept the floating matter of the air; the bent tube T′ contains caustic potash, the tube T sulphuric acid, the one intended to remove the carbonic acid and the other the aqueous vapour of the air.

The power of the electric beam to reveal the existence of anything within the experimental tube, or the impurities of the tube itself, is extraordinary. When the experiment is made in a darkened room, a tube which in ordinary daylight appears absolutely clean, is often shown by the present mode of examination to be exceedingly filthy.

The following are some of the results obtained with this arrangement :—

Nitrite of amyl.—The vapour of this liquid was in the first instance permitted to enter the experimental tube, while the beam from the electric lamp was passing through it. Curious clouds, the cause of which was then unknown, were observed to form near the place of entry, being afterwards whirled through the tube.

The tube being again exhausted, the mixed air and vapour were allowed to enter it in the dark. The slightly convergent beam of the electric light was then sent through the mixture. For a moment the tube was *optically empty*, nothing whatever being seen within it; but before a second had elapsed a shower of particles was precipitated on the beam. The cloud thus generated became denser as the light continued to act, showing at some places vivid iridescence.

The lens of the electric lamp was now placed so as to form within the tube a strongly convergent cone of rays. The tube was cleansed and again filled in dark· ness. When the light was sent through it, the precipitation upon the beam was so rapid and intense that the cone, which a moment before was invisible, flashed suddenly forth like a solid luminous spear. The effect was the same when the air and vapour were allowed to enter the tube in diffuse daylight. The cloud, however, which shone with such extraordinary radiance under the electric beam, was invisible in the ordinary light of the laboratory.

The quantity of mixed air and vapour within the experimental tube could of course be regulated at pleasure. The rapidity of the action diminished with the attenuation of the vapour. When, for example, the mercurial column associated with the experimental tube was depressed only five inches, the action was not nearly so rapid as when the tube was full. In such cases, however, it was exceedingly interesting to observe, after some seconds of waiting, a thin streamer of delicate bluish-white cloud slowly forming along the axis of the tube, and finally swelling so as to fill it.

When dry oxygen was employed to carry in the vapour, the effect was the same as that obtained with air.

When dry hydrogen was used as a vehicle, the effect was also the same.

The effect, therefore, is not due to any interaction between the vapour of the nitrite and its vehicle.

This was further demonstrated by the deportment of the vapour itself. When it was permitted to enter the experimental tube unmixed with air or any other gas, the effect was substantially the same. Hence the seat of the observed action is the vapour.

This action is not to be ascribed to heat. As regards the glass of the experimental tube, and the air within the tube, the beam employed in these experiments was perfectly cold. It had been sifted by passing it through a solution of alum, and through the thick double-convex lens of the lamp. When the unsifted beam of the lamp was employed, the effect was still the same; the obscure calorific rays did not appear to interfere with the result.

My object here being simply to point out to chemists a method of experiment which reveals a new and beautiful series of reactions, I left to them the examination of the products of decomposition. The group of atoms forming the molecule of nitrite of amyl is obviously shaken asunder by certain specific waves of the electric beam, nitric oxide and other products, of which the *nitrate* of amyl is probably one, being the result of the decomposition. The brown fumes of nitrous acid were seen mingling with the cloud within the experimental tube. The nitrate of amyl, being less volatile than the nitrite, and not being able to maintain itself in the condition of vapour, would be precipitated as a visible cloud along the track of the beam.

In the anterior portions of the tube a powerful sifting of the beam by the vapour occurs, which diminishes the chemical action in the posterior portions. In some experiments the precipitated cloud only extended half-way down the tube. When, under these circumstances, the lamp was shifted so as to send the beam through the other end of the tube, copious precipitation occurred there also.

Solar light also effects the decomposition of the nitrite-of-amyl vapour. On October 10, 1868, I partially darkened a small room in the Royal Institution, into which the sun shone, permitting the light to enter through an open portion of the window-shutter. In the track of the beam was placed a large plano-convex lens, which formed a fine convergent cone in the dust of the room behind it. The experimental tube was filled in the laboratory, covered with a black cloth, and carried into the partially darkened room. On thrusting one end of the tube into the cone of rays behind the lens, precipitation within the cone was copious and immediate. The vapour at the distant end of the tube was in part shielded by that in front, and was also more feebly acted on through the divergence of the rays. On reversing the tube, a second and similar cone was precipitated.

Physical Considerations.

I sought to determine the particular portion of the light which produced the foregoing effects. When, previous to entering the experimental tube, the beam was caused to pass through a red glass, the effect was greatly weakened, but not extinguished. This was also the case with various samples of yellow glass. A blue glass being introduced before the removal of the yellow or the red, on taking the latter away prompt precipitation occurred along the track of the blue beam. Hence, in this case, the more refrangible rays are the most chemically active. The colour of the liquid nitrite of amyl indicates that this must be the case; it is a feeble but distinct yellow: in other words, the yellow portion of the beam is most freely transmitted. It is not, however, the transmitted portion of any beam which produces chemical action, but the absorbed portion. Blue, as the complementary colour to yellow, is here absorbed, and hence the more energetic action of the blue rays.

This reasoning, however, assumes that the same rays are absorbed by the liquid and its vapour. The assumption is worth testing. A solution of the yellow chromate of potash, the colour of which may be made almost, if not altogether, identical with that of the liquid nitrite of amyl, was found far more effective in stopping the chemical rays than either the red or the yellow glass. But of all substances the liquid nitrite itself is most potent in arresting the rays which act upon its vapour. A layer one-eighth of an inch in thickness, which scarcely perceptibly affected the luminous intensity, absorbed the entire chemical energy of the concentrated beam of the electric light.

The close relation subsisting between a liquid and its vapour, as regards their action upon radiant heat, has been already amply demonstrated.[1] As regards the nitrite of amyl, this relation is more specific than in the cases hitherto adduced ; for here the special constituent of the beam, which provokes the decomposition of the vapour, is shown to be arrested by the liquid.

A question of extreme importance in molecular physics here arises : What is the real mechanism of this absorption, and where is its seat ?[2] I figure, as others do, a molecule as a group of atoms, held together by their mutual forces, but still capable of motion among themselves. The vapour of the nitrite of amyl is to

---

[1] 'Phil. Trans.' 1864 ; 'Heat, a Mode of Motion,' chap. xii. ; and p. 61 of this volume.

[2] My attention was very forcibly directed to this subject some years ago by a conversation with my excellent friend Professor Clausius.

be regarded as an assemblage of such molecules. Tho question now before us is this : In the act of absorption, is it the molecules that are effective, or is it their constituent atoms?   Is the *vis viva* of the intercepted light-waves transferred to the molecule as a whole, or to its constituent parts ?

The molecule, as a whole, can only vibrate in virtue of the forces exerted between it and its neighbour molecules.   The intensity of these forces, and consequently the rate of vibration, would, in this case, be a function of the distance between the molecules.   Now the identical absorption of the liquid and of the vaporous nitrite of amyl indicates an identical vibrating period on the part of liquid and vapour, and this, to my mind, amounts to an experimental proof that the absorption occurs in the main *within* the molecule.   For it can hardly be supposed, if the absorption were the act of the molecule as a whole, that it could continue to affect waves of the same period after the substance had passed from the vaporous to the liquid state.

In point of fact, the decomposition of the nitrite of amyl is itself to some extent an illustration of this internal molecular absorption ; for were the absorption the act of the molecule as a whole, the *relative* motions of its constituent atoms would remain unchanged, and there would be no mechanical cause for their separation. It is probably the synchronism of the vibrations of one portion of the molecule with the incident waves, that enables the amplitude of those vibrations to augment, until the chain which binds the parts of the molecule together is snapped asunder.

I anticipate wide, if not entire, generality for the fact that a liquid and its vapour absorb the same rays. A cell of liquid chlorine would, I imagine, deprive light more effectually of its power of causing chlorine and hydrogen to combine than any other filter of the luminous rays. The rays which give chlorine its colour have nothing to do with this combination, those that are absorbed by the chlorine being the really effective rays. A highly sensitive bulb, containing chlorine and hydrogen, in the exact proportions necessary for the formation of hydrochloric acid, was placed at one end of an experimental tube, the beam of the electric lamp being sent through it from the other. The bulb did not explode when the tube was filled with chlorine, while the explosion was violent and immediate when the tube was filled with air. I anticipate for the liquid chlorine an action similar to, but still more energetic than, that exhibited by the gas. If this should prove to be the case, it will favour the view that chlorine itself is *molecular* and not *monatomic*.

Production of Sky-blue by the Decomposition of Nitrite of Amyl.

When the quantity of nitrite vapour is considerable, and the light intense, the chemical action is exceedingly rapid, the particles precipitated being so large as to whiten the luminous beam. Not so, however, when a well-mixed and highly attenuated vapour fills the experimental tube. The effect now to be described was first obtained when the vapour of the nitrite was derived from a portion of its liquid which had been accidentally introduced into the passage through which the dry air flowed into the experimental tube.

In this case, the electric beam traversed the tube for several seconds before any action was visible. Decomposition then visibly commenced, and advanced slowly. When the light was very strong, the cloud appeared of a milky blue. When, on the contrary, the intensity was moderate, the blue was pure and deep. In Brücke's important experiments on the blue of the sky and the morning and evening red, pure mastic is dissolved in alcohol, and then dropped into water well stirred. When the proportion of mastic to alcohol is correct, the resin is precipitated so finely as to elude the highest microscopic power. By reflected light, such a medium appears bluish, by transmitted light yellowish, which latter colour, by augmenting the quantity of the precipitate, can be caused to pass into orange or red.

But the development of colour in the attenuated nitrite-of-amyl vapour is doubtless more similar to what takes place in our atmosphere. The blue, moreover, is far purer and more sky-like than that obtained from Brücke's turbid medium. Never, even in the skies of the Alps, have I seen a richer or a purer blue than that attainable by a suitable disposition of the light falling upon the precipitated vapour.

Iodide of Allyl.—Among the liquids hitherto subjected to the concentrated electric light, iodide of allyl, in point of rapidity and intensity of action, comes next to the nitrite of amyl. With the iodide I have employed both oxygen and hydrogen, as well as air, as a vehicle, and found the effect in all cases substantially the same. The cloud-column here was exquisitely beautiful. It revolved round the axis of the decomposing beam; it was nipped at certain places like an hour-glass, and round the two bells of the glass delicate cloud-filaments twisted themselves in spirals. It also folded itself into convolutions resembling those of shells. In certain conditions of the atmosphere in the Alps I have often observed clouds of a special pearly lustre; when hydrogen was made the vehicle of the iodide-of-allyl vapour a similar lustre was most exquisitely shown.

With a suitable disposition of the light, the purple hue of iodine-vapour came out very strongly in the tube.

The remark already made, as to the bearing of the decomposition of nitrite of amyl by light on the question of molecular absorption, applies here also ; for were the absorption the work of the molecule as a whole, the iodine would not be dislodged from the allyl with which it is combined. The non-synchronism of iodine with the waves of obscure heat is illustrated by its marvellous transparency to such heat. May not its synchronism with the waves of light in the present instance be the cause of its divorce from the allyl ?

Iodide of Isopropyl.—The action of light upon the vapour of this liquid is, at first, more languid than upon iodide of allyl ; indeed many beautiful reactions may be overlooked, in consequence of this languor at the commencement. After some minutes' exposure, however, clouds begin to form, which grow in density and in beauty as the light continues to act. In every experiment hitherto made with this substance the column of cloud filling the experimental tube, was divided into two distinct parts near the middle of the tube. In one experiment a globe of cloud formed at the centre, from which, right and left, issued an axis uniting the globe with two adjacent cylinders. Both globe and cylinders were animated by a common motion of rotation. As the action continued, paroxysms of motion were manifested ; the various parts of the cloud would rush through each other with sudden violence. During these motions beautiful and grotesque cloud-forms were developed. At some places the nebulous mass would become ribbed so as to resemble the graining of wood ; a longitudinal motion would at times generate in it a series of curved transverse bands, the retarding influence of the sides of the tube causing an appearance resembling, on a small scale, the dirt-bands of the Mer de Glace.   In the anterior portion of the tube those sudden commotions were most intense; here buds of cloud would sprout forth, and grow in a few seconds into perfect flower-like forms.   The cloud of iodide of isopropyl had a character of its own, and differed materially from all others that I had seen.   A gorgeous mauve colour was observed in the last twelve inches of the tube ; the vapour of iodine was present, and it may have been the sky-blue scattered by the precipitated particles which, mingling with the purple of the iodine, produced the mauve.   As in all other cases here adduced, the effects were proved to be due to the light ; they never occurred in darkness.

The forms assumed by some of those *actinic clouds*, as I propose to call them, in consequence of rotations and other motions, due to differences of temperature, are perfectly astounding.   I content myself here with a meagre description of one more of them.

The tube being filled with the sensitive mixture, the beam was sent through it, the lens at the same time being so placed as to produce a cone of very intense light.   Two minutes elapsed before anything was visible; but at the end of this time a faint bluish cloud appeared to hang itself on the most concentrated portion of the beam.

Soon afterwards a second cloud was formed five inches farther down the experimental tube.   Both clouds were united by a slender cord of the same bluish tint as themselves.

As the action of the light continued, the first cloud gradually resolved itself into a series of parallel disks of exquisite delicacy, which rotated round an axis perpendicular to their surfaces, and finally blended to a screw surface with an inclined generatrix.   This gradually changed into a filmy funnel, from the narrow end of which the ' cord ' extended to the cloud in advance.

The latter also underwent slow but incessant modification. It first resolved itself into a series of strata resembling those of the electric discharge. After a little time, and through changes which it was difficult to follow, both clouds presented the appearance of a series of concentric funnels set one within the other, the interior ones being seen through the outer ones. Those of the distant cloud resembled claret-glasses in shape. As many as six funnels were thus concentrically set together, the two series being united by the delicate cord of cloud already referred to. Other cords and slender tubes were afterwards formed, which coiled themselves in delicate spirals around the funnels.

Rendering the light along the connecting-cord more intense, it diminished in thickness and became whiter; this was a consequence of the enlargement of its particles. The cord finally disappeared, while the funnels melted into two ghost-like films, shaped like parasols. They were barely visible, being of an exceedingly delicate blue tint. They seemed woven of blue air. To compare them with cobweb or with gauze would be to liken them to something infinitely grosser than themselves.

In all cases a distant candle-flame, when looked at through the cloud, was sensibly undimmed.

§ 2. ON THE BLUE COLOUR OF THE SKY, AND THE POLARISATION OF SKYLIGHT.[1]

1869.

After the communication to the Royal Society of the foregoing brief account of a new Series of Chemical Reactions produced by Light, the experiments upon

[1] In my 'Lectures on Light' (Longmans), the polarisation of light will be found briefly, but, I trust, clearly explained.

this subject were continued, the number of substances thus acted on being considerably increased.

I now, however, beg to direct attention to two questions glanced at incidentally in the preceding pages—the blue colour of the sky, and the polarisation of skylight. Reserving the historic treatment of the subject for a more fitting occasion, I would merely mention now that these questions constitute, in the opinion of our most eminent authorities, the two great standing enigmas of meteorology. Indeed it was the interest manifested in them by Sir John Herschel, in a letter of singular speculative power, addressed to myself, that caused me to enter upon the consideration of these questions so soon.

The apparatus with which I work consists, as already stated, of a glass tube about a yard in length, and from $2\frac{1}{2}$ to 3 inches internal diameter. The vapour to be examined is introduced into this tube in the manner already described, and upon it the condensed beam of the electric lamp is permitted to act, until the neutrality or the activity of the substance has been declared.

It has hitherto been my aim to render the chemical action of light upon vapours visible. For this purpose substances have been chosen, one at least of whose products of decomposition under light shall have a boiling-point so high, that as soon as the substance is formed it shall be precipitated. By graduating the quantity of the vapour, this precipitation may be rendered of any degree of fineness, forming particles distinguishable by the naked eye, or far beyond the reach of our highest microscopic powers. I have no reason to doubt that particles may be thus obtained, whose diameters constitute but a small fraction of the length of a wave of violet light.

In all cases when the vapours of the liquids employed are sufficiently attenuated, no matter what the liquid may be, the visible action commences with the formation of a *blue cloud*. But here I must guard myself against all misconception as to the use of this term. The 'cloud' here referred to is totally invisible in ordinary daylight. To be seen, it requires to be surrounded by darkness, *it only* being illuminated by a powerful beam of light. This blue cloud differs in many important particulars from the finest ordinary clouds, and might justly have assigned to it an intermediate position between such clouds and true vapour. With this explanation, the term 'cloud,' or 'incipient cloud,' or 'actinic cloud,' as I propose to employ it, cannot, I think, be misunderstood.

I had been endeavouring to decompose carbonic acid gas by light. A faint bluish cloud, due it may be, or it may not be, to the residue of some vapour previously employed, was formed in the experimental tube On looking across this cloud through a Nicol's prism, the line of vision being horizontal, it was found that when the short diagonal of the prism was vertical, the quantity of light reaching the eye was greater than when the long diagonal was vertical. When a plate of tourmaline was held between the eye and the bluish cloud, the quantity of light reaching the eye when the axis of the prism was perpendicular to the axis of the illuminating beam, was greater than when the axes of the crystal and of the beam were parallel to each other.

This was the result all round the experimental tube. Causing the crystal of tourmaline to revolve round the tube, with its axis perpendicular to the illuminating beam, the quantity of light that reached the eye was in all its positions a maximum. When the crystallographic axis was parallel to the axis of the beam, the quantity of light transmitted by the crystal was a minimum.

From the illuminated bluish cloud, therefore, polarised light was discharged, the direction of maximum polarisation being at right angles to the illuminating beam; the plane of vibration of the polarised light was perpendicular to the beam.[1]

Thin plates of selenite or of quartz, placed between the Nicol and the actinic cloud, displayed the colours of polarised light, these colours being most vivid when the line of vision was at right angles to the experimental tube. The plate of selenite usually employed was a circle, thinnest at the centre, and augmenting uniformly in thickness from the centre outwards. When placed in its proper position between the Nicol and the cloud, it exhibited a system of splendidly-coloured rings.

The cloud here referred to was the first operated upon in the manner described. It may, however, be greatly improved upon by the choice of proper substances, and by the application, in proper quantities, of the substances chosen. Benzol, bisulphide of carbon, nitrite of amyl, nitrite of butyl, iodide of allyl, iodide of isopropyl, and many other substances may be employed. I will take the nitrite of butyl as illustrative of the means adopted to secure the best result, with reference to the present question.

And here it may be mentioned that a vapour, which when alone, or mixed with air in the experimental tube, resists the action of light, or shows but a feeble result of this action, may, when placed in proximity with another gas or vapour, exhibit vigorous, if not violent action. The case is similar to that of carbonic acid gas, which, diffused in the atmosphere, resists the de-

---

[1] This is still an undecided point; but the probabilities are so much in its favour, and it is in my opinion so much preferable to have a physical image on which the mind can rest, that I do not hesitate to employ the phraseology in the text.

composing action of solar light, but when placed in contiguity with chlorophyl in the leaves of plants, has its molecules shaken asunder.

Dry air was permitted to bubble through the liquid nitrite of butyl, until the experimental tube, which had been previously exhausted, was filled with the mixed air and vapour. The visible action of light upon the mixture after fifteen minutes' exposure was slight. The tube was afterwards filled with half an atmosphere of the mixed air and vapour, and a second half-atmosphere of air which had been permitted to bubble through fresh commercial hydrochloric acid. On sending the beam through this mixture, the tube, for a moment, was optically empty. But the pause amounted only to a small fraction of a second, a dense cloud being immediately precipitated upon the beam.

This cloud began blue, but the advance to whiteness was so rapid as almost to justify the application of the term 'instantaneous. The dense cloud, looked at perpendicularly to its axis, showed scarcely any signs of polarisation. Looked at obliquely the polarisation was strong.

The experimental tube being again cleansed and exhausted, the mixed air and nitrite-of-butyl vapour was permitted to enter it until the associated mercury column was depressed $\frac{1}{10}$ of an inch. In other words, the air and vapour, united, exercised a pressure not exceeding $\frac{1}{300}$ of an atmosphere. Air, passed through a solution of hydrochloric acid, was then added, till the mercury column was depressed three inches. The condensed beam of the electric light was passed for some time through this mixture without revealing anything within the tube competent to scatter the light. Soon, however, a superbly blue cloud was formed along the track of the beam, and it continued blue sufficiently long to permit of its thorough examination. The light discharged from the cloud, at right angles to its own length, was at first perfectly polarised. It could be totally quenched by the Nicol. By degrees the cloud became of whitish blue, and for a time the selenite colours, obtained by looking at it normally, were exceedingly brilliant. The direction of maximum polarisation was distinctly at right angles to the illuminating beam. This continued to be the case as long as the cloud maintained a decided blue colour, and even for some time after the blue had changed to whitish blue. But, as the light continued to act, the cloud became coarser and whiter, particularly at its centre, where it at length ceased to discharge polarised light in the direction of the perpendicular, while it continued to do so at both ends.

But the cloud which had thus ceased to polarise the light emitted normally, showed vivid selenite colours when looked at obliquely, proving that the direction of maximum polarisation changed with the texture of the cloud. This point shall receive further illustration subsequently.

A blue, equally rich and more durable, was obtained by employing the nitrite-of-butyl vapour in a still more attenuated condition. The instance here cited is representative. In all cases, and with all substances, the cloud formed at the commencement, when the precipitated particles are sufficiently fine, is *blue*, and it can be made to display a colour rivalling that of the purest Italian sky. In all cases, moreover, this fine blue cloud polarises *perfectly* the beam which illuminates it, the direction of polarisation enclosing an angle of 90° with the axis of the illuminating beam.

It is exceedingly interesting to observe both the perfection and the decay of this polarisation. For ten or fifteen minutes after its first appearance the light from a vividly illuminated actinic cloud, looked at perpendicularly, is absolutely quenched by a Nicol's prism with its longer diagonal vertical. But as the sky-blue is gradually rendered impure by the growth of the particles—in other words, as real clouds begin to be formed—the polarisation begins to decay, a portion of the light passing through the prism in all its positions. It is worthy of note, that for some time after the cessation of perfect polarisation, the residual light which passes, when the Nicol is in its position of minimum transmission, is of a gorgeous blue, the whiter light of the cloud being extinguished.[1] When the cloud texture has become sufficiently coarse to approximate to that of ordinary clouds, the rotation of the Nicol ceases to have any sensible effect on the quantity of light discharged normally.

The perfection of the polarisation, in a direction perpendicular to the illuminating beam, is also illustrated by the following experiment: A Nicol's prism, large enough to embrace the entire beam of the electric lamp, was placed between the lamp and the experimental tube. A few bubbles of air, carried through the liquid nitrite of butyl, were introduced into the tube, and they were followed by about three inches (measured by the mercurial gauge) of air which had passed through aqueous hydrochloric acid. Sending the polarised beam through the tube, I placed myself in front of it, my eye being on a level with its axis, my assistant occupying a similar position behind the tube. The short diagonal of the large Nicol was in the first instance vertical, the plane of vibration of the emergent

---

[1] This shows that particles too large to polarise the blue, polarise perfectly light of lower refrangibility.

beam being therefore also vertical. As the light continued to act, a superb blue cloud, visible to both my assistant and myself, was slowly formed. But this cloud, so deep and rich when looked at from the positions mentioned, *utterly disappeared when looked at vertically downwards, or vertically upwards.* Reflection from the cloud was not possible in these directions. When the large Nicol was slowly turned round its axis, the eye of the observer being on the level of the beam, and the line of vision perpendicular to it, entire extinction of the light emitted horizontally occurred when the longer diagonal of the large Nicol was vertical. But now a vivid blue cloud was seen when looked at downwards or upwards. This truly fine experiment, which I contemplated making on my own account, was first definitely suggested by a remark in a letter addressed to me by Professor Stokes.

As regards the polarisation of skylight, the greatest stumbling-block has hitherto been, that, in accordance with the law of Brewster, which makes the index of refraction the tangent of the polarising angle, the reflection which produces perfect polarisation would require to be made *in* air *upon* air; and indeed this led many of our most eminent men, Brewster himself among the number, to entertain the idea of aërial molecular reflection.[1]   I have, however, operated upon

[1] 'The cause of the polarisation is evidently a reflection of the sun's light upon *something*. The question is on what? Were the angle of maximum polarisation 76°, we should look to water or ice as the reflecting body, however inconceivable the existence in a cloudless atmosphere and a hot summer's day of unevaporated molecules (particles?) of water. But though we were once of this opinion, careful observation has satisfied us that 90°, or thereabouts, is the correct angle, and that therefore whatever be the body on which the light has been reflected, *if polarised by a single reflection,* the polarising angle must be 45°, and the index of refraction, which is the tangent of that angle, unity; in other words, the

substances of widely different refractive indices, and therefore of very different polarising angles as ordinarily defined, but the polarisation of the beam, by the incipient cloud, has thus far proved itself to be absolutely independent of the polarising angle. The law of Brewster does not apply to matter in this condition, and it rests with the undulatory theory to explain why. Whenever the precipitated particles are sufficiently fine, no matter what the substance forming the particles may be, the direction of maximum polarisation is at right angles to the illuminating beam, the polarising angle for matter in this condition being invariably 45°.

Suppose our atmosphere surrounded by an envelope impervious to light, but with an aperture on the sunward side through which a parallel beam of solar light could enter and traverse the atmosphere. Surrounded by air not directly illuminated, the track of such a beam would resemble that of the parallel beam of the electric lamp through an incipient cloud. The sunbeam would be blue, and it would discharge laterally light in precisely the same condition as that discharged by the incipient cloud. In fact, the azure revealed by such a beam would be to all intents and purposes that which I have called a 'blue cloud.' Conversely our 'blue cloud' is, to all intents and purposes, an *artificial sky.*[1]

reflection would require to be made *in* air *upon* air !' (Sir John Herschel, 'Meteorology,' par. 233.)

Any particles, if small enough, will produce both the colour and the polarisation of the sky. But is the existence of small water-particles on a hot summer's day *in the higher regions of our atmosphere* inconceivable? It is to be remembered that the oxygen and nitrogen of the air behave as a vacuum to radiant heat, the exceedingly attenuated vapour of the higher atmosphere being therefore in practical contact with the cold of space.

[1] The opinion of Sir John Herschel, connecting the polarisation and the blue colour of the sky, is verified by the foregoing results.

But, as regards the polarisation of the sky, we know that not only is the direction of maximum polarisation at right angles to the track of the solar beams, but that at certain angular distances, probably variable ones, from the sun, 'neutral points,' or points of no polarisation, exist, on both sides of which the planes of atmospheric polarisation are at right angles to each other. I have made various observations upon this subject which are reserved for the present; but, pending the more complete examination of the question, the following facts bearing upon it may be submitted.

The parallel beam employed in these experiments tracked its way through the laboratory air, exactly as sunbeams are seen to do in the dusty air of London. I have reason to believe that a great portion of the matter thus floating in the laboratory air consists of organic particles, which are capable of imparting a perceptibly bluish tint to the air. These also showed, though far less vividly, all the effects of polarisation obtained with the incipient clouds. The light discharged laterally from the track of the illuminating beam was polarised, though not perfectly, the direction of maximum polarisation being at right angles to the beam. At all points of the beam, moreover, throughout its entire length, the light emitted normally was in the same state of polarisation. Keeping the positions of the Nicol and the selenite constant, the same colours were observed

'The more the subject [the polarisation of skylight] is considered,' writes this eminent philosopher, 'the more it will be found beset with difficulties, and its explanation when arrived at will probably be found to carry with it that of the blue colour of the sky itself, and of the great quantity of light it actually does send down to us.' 'We may observe, too,' he adds, 'that it is only where the purity of the sky is most absolute that the polarisation is developed in its highest degree, and that where there is the slightest perceptible tendency to cirrus it is materially impaired.' This applies word for word to our 'incipient clouds.'

throughout the entire beam, when the line of vision was perpendicular to its length.

The horizontal column of air, thus illuminated, was 18 feet long, and could therefore be looked at very obliquely. I placed myself near the end of the beam, as it issued from the electric lamp, and, looking through the Nicol and selenite more and more obliquely at the beam, observed the colours fading until they disappeared. Augmenting the obliquity the colours appeared once more, but they were now complementary to the former ones.

Hence this beam, like the sky, exhibited a neutral point, on opposite sides of which the light was polarised in planes at right angles to each other.

Thinking that the action observed in the laboratory might be caused, in some way, by the vaporous fumes diffused in its air, I had the light removed to a room at the top of the Royal Institution. The track of the beam was seen very finely in the air of this room, a length of 14 or 15 feet being attainable. This beam exhibited all the effects observed with the beam in the laboratory. Even the uncondensed electric light falling on the floating matter showed, though faintly, the effects of polarisation.

When the air was so sifted as to entirely remove the visible floating matter, it no longer exerted any sensible action upon the light, but behaved like a vacuum. The light is scattered and polarised by *particles*, not by molecules or atoms.

By operating upon the fumes of chloride of ammonium, the smoke of brown paper, and tobacco-smoke, I had varied and confirmed in many ways those experiments on neutral points, when my attention was drawn by Sir Charles Wheatstone to an important observation communicated to the Paris Academy in 1860 by Pro-

9

fessor Govi, of Turin.[1]   M. Govi had been led to examine
a beam of light sent through a room in which were successively diffused the smoke of incense, and tobacco-
smoke.   His first brief communication stated the fact
of polarisation by such smoke ; but in his second communication he announced the discovery of a neutral
point in the beam, at the opposite sides of which the
light was polarised in planes at right angles to each
other.

But unlike my observations on the laboratory air,
and unlike the action of the sky, the direction of maximum polarisation in M. Govi's experiment enclosed a
very small angle with the axis of the illuminating beam.
The question was left in this condition, and I am not
aware that M. Govi or any other investigator has pursued it further.

I had noticed, as before stated, that as the clouds
formed in the experimental tube became denser, the
polarisation of the light discharged at right angles to
the beam became weaker, the direction of maximum
polarisation becoming oblique to the beam.   Experiments on the fumes of chloride of ammonium gave me
also reason to suspect that the position of the neutral
point was not constant, but that it varied with the
density of the illuminated fumes.

The examination of these questions led to the following new and remarkable results : The laboratory being
well filled with the fumes of incense, and sufficient time
being allowed for their uniform diffusion, the electric
beam was sent through the smoke.   From the track of
the beam polarised light was discharged ; but the direction of maximum polarisation, instead of being perpendicular, now enclosed an angle of only 12° or 13° with
the axis of the beam.

[1] 'Comptes Rendus,' tome li. pp. 360 and 669.

A neutral point, with complementary effects at opposite sides of it, was also exhibited by the beam. The angle enclosed by the axis of the beam, and a line drawn from the neutral point to the observer's eye, measured in the first instance 66°.

The windows of the laboratory were now opened for some minutes, a portion of the incense-smoke being permitted to escape. On again darkening the room and turning on the light, the line of vision to the neutral point was found to enclose, with the axis of the beam, an angle of 63°.

The windows were again opened for a few minutes, more of the smoke being permitted to escape. Measured as before, the angle referred to was found to be 54°.

This process was repeated three additional times; the neutral point was found to recede lower and lower down the beam, the angle between a line drawn from the eye to the neutral point and the axis of the beam falling successively from 54° to 49°, 43° and 33°.

The distances, roughly measured, of the neutral point from the lamp, corresponding to the foregoing series of observations, were these :—

1st observation	.	.	.	2 feet	2 inches.
2nd ,,	.	.	.	2 ,,	6 ,,
3rd ,,	.	.	.	2 ,,	10 ,,
4th ,,	.	.	.	3 ,,	2 ,,
5th ,,	.	.	.	3 ,,	7 ,,
6th ,,	.	.	.	4 ,,	6 ,,

At the end of this series of experiments the direction of maximum polarisation had again become normal to the beam.

The laboratory was next filled with the fumes of gunpowder. In five successive experiments, corresponding to five different densities of the gunpowder-smoke, the angles enclosed between the line of vision to the neutral point and the axis of the beam, were 63°, 50°, 47°, 42°, and 38° respectively.

After the clouds of gunpowder had cleared away, the laboratory was filled with the fumes of common resin, rendered so dense as to be very irritating to the lungs. The direction of maximum polarisation enclosed, in this case, an angle of 12°, or thereabouts, with the axis of the beam. Looked at, as in the former instances, from a position near the electric lamp, no neutral point was observed throughout the entire extent of the beam.

When this beam was looked at normally through the selenite and Nicol, the ring-system, though not brilliant, was distinct. Keeping the eye upon the plate of selenite, and the line of vision perpendicular, the windows were opened, the blinds remaining undrawn. The resinous fumes slowly diminished, and as they did so the ring-system became paler. It finally disappeared. Continuing to look in the same direction, the rings revived, but now the colours were complementary to the former ones. *The neutral point had passed me in its motion down the beam, consequent upon the attenuation of the fumes of resin.*

With the fumes of chloride of ammonium substantially the same results were obtained. Sufficient, however, has been here stated to illustrate the variability of the position of the neutral point.[1]

By a puff of tobacco-smoke, or of condensed steam, blown into the illuminated beam, the brilliancy of the selenite colours may be greatly enhanced. But with different clouds two different effects are produced. Let the ring-system observed in the common air be brought to its maximum strength, and then let an attenuated

[1] Brewster has proved the variability of the position of the neutral point for skylight with the sun's altitude, a result obviously connected with the foregoing experiments.

cloud of chloride of ammonium be thrown into the beam at the point looked at; the ring system flashes out with augmented brilliancy, but the character of the polarisation remains unchanged. This is also the case when phosphorus, or sulphur, is burned underneath the beam, so as to cause the fine particles of phosphorus or of sulphur to rise into the light. With the sulphur-fumes the brilliancy of the colours is exceedingly intensified; but in none of these cases is there any change in the character of the polarisation.

But when a puff of the fumes of hydrochloric acid, hydriodic acid, or nitric acid is thrown into the beam, there is a complete reversal of the selenite tints. Each of these clouds twists the plane of polarisation 90°, causing the centre of the ring-system to change from black to white, and the rings themselves to emit their complementary colours.[1]

Almost all liquids have motes in them sufficiently numerous to polarise sensibly the light, and very beautiful effects may be obtained by simple artificial devices. When, for example, a cell of distilled water is placed in front of the electric lamp, and a thin slice of the beam is permitted to pass through it, scarcely any polarised light is discharged, and scarcely any colour produced with a plate of selenite. But if a bit of soap be agitated in the water above the beam, the moment the infinitesimal particles reach the light the liquid sends forth laterally almost perfectly polarised light; and if the selenite be employed, vivid colours flash into existence. A still more brilliant result is

---

[1] Sir John Herschel suggested to me that this change of the polarisation from positive to negative may indicate a change from polarisation by reflection to polarisation by refraction. This thought repeatedly occurred to me while looking at the effects; but it will require much following up before it emerges into clearness.

obtained with mastic dissolved in a great excess of
alcohol.

The selenite rings, in fact, constitute an extremely
delicate test as to the collective quantity of individually
invisible particles in a liquid. Commencing with distilled water, for example, a thick slice of light is necessary to make the polarisation of its suspended particles
sensible. A much thinner slice suffices for common
water ; while, with Brücke's precipitated mastic, a
slice too thin to produce any sensible effect with most
other liquids, suffices to bring out vividly the selenite
colours.

§ 3.   The Sky of the Alps.

The vision of an object always implies a differential
action on the retina of the observer. The object is
distinguished from surrounding space by its excess or
defect of light in relation to that space. By altering
the illumination, either of the object itself or of its
environment, we alter the appearance of the object.
Take the case of clouds floating in the atmosphere with
patches of blue between them. Anything that changes
the illumination of either alters the appearance of both,
that appearance depending, as stated, upon differential
action. Now the light of the sky, being polarised,
may, as the reader of the foregoing pages knows, be in
great part quenched by a Nicol's prism, while the light
of a common cloud, being unpolarised, cannot be thus
extinguished. Hence the possibility of very remarkable
variations, not only in the aspect of the firmament,
which is really changed, but also in the aspect of the
clouds, which have that firmament as a background.
It is possible, for example, to choose clouds of such a
depth of shade that when the Nicol quenches the light
behind them, they shall vanish, being undistinguishable

from the residual dull tint which outlives the extinction
of the brilliancy of the sky. A cloud less deeply
shaded, but still deep enough, when viewed with the
naked eye, to appear dark on a bright ground, is suddenly changed to a white cloud on a dark ground by,
the quenching of the light behind it. When a reddish
cloud at sunset chances to float in the region of maximum polarisation, the quenching of the surrounding
light causes it to flash with a brighter crimson. Last
Easter eve the Dartmoor sky, which had just been
cleansed by a snow-storm, wore a very wild appearance.
Round the horizon it was of steely brilliancy, while
reddish cumuli and cirri floated southwards. When the
sky was quenched behind them these floating masses
seemed like dull embers suddenly blown upon; they
brightened like a fire.

In the Alps we have the most magnificent examples
of crimson clouds and snows, so that the effects just
referred to may be here studied under the best possible
conditions. On August 23, 1869, the evening Alpen-
glow was very fine, though it did not reach its maximum
depth and splendour. The side of the Weisshorn seen
from the Bel Alp, being turned from the sun, was tinted
mauve; but I wished to observe one of the rose-coloured
buttresses of the mountain. Such a one was visible
from a point a few hundred feet above the hotel. The
Matterhorn also, though for the most part in shade,
had a crimson projection, while a deep ruddy red
lingered along its western shoulder. Four distinct
peaks and buttresses of the Dom, in addition to its
dominant head—all covered with pure snow—were
reddened by the light of sunset. The shoulder of the
Alphubel was similarly coloured, while the great mass
of the Fletschorn was all a-glow, and so was the snowy
spine of the Monte Leone.

Looking at the Weisshorn through the Nicol, the glow of its protuberance was strong or weak according to the position of the prism. The summit also underwent striking changes. In one position of the prism it exhibited a pale white against a dark background; in the rectangular position it was a dark mauve against a light background. The red of the Matterhorn changed in a similar manner; but the whole mountain also passed through wonderful changes of definition. The air at the time was filled with a silvery haze, in which the Matterhorn almost disappeared. This could be wholly quenched by the Nicol, and then the mountain sprang forth with astonishing solidity and detachment from the surrounding air. The changes of the Dom were still more wonderful. A vast amount of light could be removed from the sky behind it, for it occupied the position of maximum polarisation. By a little practice with the Nicol it was easy to render the extinction of the light, or its restoration, almost instantaneous. When the sky was quenched, the four minor peaks and buttresses, and the summit of the Dom, together with the shoulder of the Alphubel, glowed as if set suddenly on fire. This was immediately dimmed by turning the Nicol through an angle of 90°. It was not the stoppage of the light of the sky behind the mountains alone which produced this startling effect; the air between them and me was highly opalescent, and the quenching of this intermediate glare augmented remarkably the distinctness of the mountains.

On the morning of August 24 similar effects were finely shown. At 10 A.M. all three mountains, the Dom, the Matterhorn, and the Weisshorn, were powerfully affected by the Nicol. But in this instance also, the line drawn to the Dom being very nearly perpendicular to the solar beams, the effects on this mountain were most striking. The grey summit of the Matterhorn, at the same time, could scarcely be distinguished from the opalescent haze around it; but when the Nicol quenched the haze, the summit became instantly isolated, and stood out in bold definition. It is to be remembered that in the production of these effects the only things changed are the sky behind, and the luminous haze in front of the mountains; that these are changed because the light emitted from the sky and from the haze is plane polarised light, and that the light from the snows and from the mountains, being sensibly unpolarised, is not directly affected by the Nicol. It will also be understood that it is not the interposition of the haze *as an opaque body* that renders the mountains indistinct, but that it is the *light* of the haze which dims and bewilders the eye, and thus weakens the definition of objects seen through it.

These results have a direct bearing upon what artists call ' aërial perspective.' As we look from the summit of Mont Blanc, or from a lower elevation, at the serried crowd of peaks, especially if the mountains be darkly coloured—covered with pines, for example — every peak and ridge is separated from the mountains behind it by a thin blue haze which renders the relations of the mountains as to distance unmistakable. When this haze is regarded through the Nicol perpendicular to the sun's rays, it is in many cases wholly quenched, because the light which it emits in this direction is wholly polarised. When this happens, aërial perspective is abolished, and mountains very differently distant appear to rise in the same vertical plane. Close to the Bel Alp, for instance, is the gorge of the Massa, and beyond the gorge is a high ridge darkened by pines. This ridge may be projected upon the dark slopes at the opposite side of the Rhone valley, and between both we have the blue haze referred to, throwing the distant mountains far away.   But at certain hours of the day the haze may be quenched, and then the Massa ridge and the mountains beyond the Rhone seem almost equally distant from the eye.   The one appears, as it were, a vertical continuation of the other.   The haze varies with the temperature and humidity of the atmosphere.   At certain times and places it is almost as blue as the sky itself; but to see its colour, the attention must be withdrawn from the mountains and from the trees which cover them.   In point of fact, the haze is a piece of more or less perfect sky; it is produced in the same manner, and is subject to the same laws, as the firmament itself.   We live *in* the sky, not *under* it.

These points were further elucidated by the deportment of the selenite plate, with which the readers of the foregoing pages are so well acquainted.   On some of the sunny days of August the haze in the valley of the Rhone, as looked at from the Bel Alp, was very remarkable.   Towards evening the sky above the mountains opposite to my place of observation yielded a series of the most splendidly-coloured iris-rings; but on lowering the selenite until it had the darkness of the pines at the opposite side of the Rhone valley, instead of the darkness of space, as a background, the colours were not much diminished in brilliancy.   I should estimate the distance across the valley, as the crow flies, to the opposite mountain, at nine miles; so that a body of air of this thickness can, under favourable circumstances, produce chromatic effects of polarisation almost as vivid as those produced by the sky itself.

Again : the light of a landscape, as of most other things, consists of two parts; the one, coming purely from superficial reflection, is always of the same colour as the light which falls upon the landscape ; the other part reaches us from a certain depth within the objects which compose the landscape, and it is this portion of the total light which gives these objects their distinctive colours. The white light of the sun enters all substances to a certain depth, and is partly ejected by internal reflection ; each distinct substance absorbing and reflecting the light, in accordance with the laws of its own molecular constitution. Thus the solar light is *sifted* by the landscape, which appears in such colours and variations of colour as, after the sifting process, reach the observer's eye. Thus the bright green of grass, or the darker colour of the pine, never comes to us alone, but is always mingled with an amount of light derived from superficial reflection. A certain hard brilliancy is conferred upon the woods and meadows by this superficially-reflected light. Under certain circumstances, it may be quenched by a Nicol's prism, and we then obtain the true colour of the grass and foliage. Trees and meadows, thus regarded, exhibit a richness and softness of tint which they never show as long as the superficial light is permitted to mingle with the true interior emission. The needles of the pines show this effect very well, large-leaved trees still better ; while a glimmering field of maize exhibits the most extraordinary variations when looked at through the rotating Nicol.

Thoughts and questions like those here referred to took me, in August 1869, to the top of the Aletschhorn. The effects described in the foregoing paragraphs were for the most part reproduced on the summit of the mountain. I scanned the whole of the sky with my Nicol. Both alone, and in conjunction with the selenite, it pronounced the perpendicular to the solar beams to be the direction of maximum polarisation.

But at no portion of the firmament was the polarisation complete. The artificial sky produced in the experiments recorded in the preceding pages could, in this respect, be rendered far more perfect than the natural one; while the gorgeous 'residual blue' which makes its appearance when the polarisation of the artificial sky ceases to be perfect, was strongly contrasted with the lack-lustre hue which, in the case of the firmament, outlived the extinction of the brilliancy. With certain substances, however, artificially treated, this dull residue may also be obtained.

All along the arc from the Matterhorn to Mont Blanc the light of the sky immediately above the mountains was powerfully acted upon by the Nicol. In some cases the variations of intensity were astonishing. I have already said that a little practice enables the observer to shift the Nicol from one position to another so rapidly as to render the alternative extinction and restoration of the light immediate. When this was done along the arc to which I have referred, the alternations of light and darkness resembled the play of sheet lightning behind the mountains. There was an element of awe connected with the suddenness with which the mighty masses, ranged along the line referred to, changed their aspect and definition under the operation of the prism.

[In the last edition of the 'Fragments of Science' an essay on 'Dust and Disease' followed here ; but as almost all my writings on the 'Germ Theory' are now collected in a single volume entitled 'Essays on the Floating Matter of the Air,' 'Dust and Disease' no longer appears in the 'Fragments.' In its place I venture to introduce a short article written early last year for an important American magazine.]

V.

THE SKY.[1]

INVITED to write for the 'Forum' an article that would have brought me face to face with ' problems of life and mind ' for which I was at the moment unprepared, and unwilling to decline a request so courteously made, I offered, if the editor cared to accept it, to send him a contribution on the subject here presented.

I mentioned this subject, thinking that, in addition to its interest as a fragment of ' natural knowledge,' it might permit of a glance at the workings of the scientific mind when engaged on the deeper problems which come before it. In the house of Science are many mansions, occupied by tenants of diverse kinds. Some of them execute with painstaking fidelity the useful work of observation, recording from day to day the aspects of Nature, or the indications of instruments devised to reveal her ways. Others there are who add to this capacity for observation a power over the language of experiment, by means of which they put questions to Nature, and receive from her intelligible replies. There is, again, a third class of minds, that cannot rest content with observation and experiment,

[1] From 'The Forum,' February 1888.

whose love of causal unity tempts them perpetually to break through the limitations of the senses, and to seek beyond them the roots and reasons of the phenomena which the observer and experimenter record. To such spirits—adventurous and firm—we are indebted for our deeper knowledge of the methods by which the physical universe is ordered and ruled.

In his efforts to cross the common bourne of the known and the unknown, the effective force of the man of science must depend, to a great extent, upon his acquired knowledge. But knowledge alone will not do; a stored memory will not suffice; inspiration must lend its aid. Scientific inspiration, however, is usually, if not always, the fruit of long reflection—of patiently 'intending the mind,' as Newton phrased it; and as Copernicus, Newton, and Darwin practised it; until outer darkness yields a glimmer, which in due time opens out into perfect intellectual day. From some of his expressions it might be inferred that Newton scorned hypotheses; but he allowed them, nevertheless, an open avenue to his own mind. He propounded the famous corpuscular theory of light, illustrating it and defending it with a skill, power, and fascination which subsequently won for it ardent supporters among the best intellects of the world. This theory, moreover, was weighted with a supplementary hypothesis, which ascribed to the luminiferous molecules 'fits of easy reflection and transmission,' in virtue of which they were sometimes repelled from the surfaces of bodies and sometimes permitted to pass through. Newton may have scorned the levity with which hypotheses are sometimes framed; but he lived in an atmosphere of theory, which he, like all profound scientific thinkers, found to be the very breath of his intellectual life.

The theorist takes his conceptions from the world

of fact, and refines and alters them to suit his needs. The sensation of sound was known to be produced by aërial waves impinging on the auditory nerve. Air being a thing that could be felt, and its vibrations, by suitable treatment, made manifest to the eye, there was here a physical basis for the ' scientific imagination ' to build upon. Both Hooke and Huyghens built upon it with effect. By the illustrious astronomer last named the conception of waves was definitely transplanted from its terrestrial birthplace to a universal medium whose undulations could only be intellectually discerned. Huyghens did not establish the undulatory theory, but he took the first firm step towards establishing it. Laying this theory at the root of the phenomena of light, he went a good way towards showing that these phenomena are the necessary outgrowth of the conception.

By analysis and synthesis Newton proved the white light of the sun to be a skein of many colours. The cause of colour was a question which immediately occupied his thoughts ; and here, as in other cases, he freely resorted to hypothesis. He saw, with his mind's eye, his luminiferous corpuscles crossing the bodily eye, and imparting successive shocks to the retina behind. To differences of ' bigness ' in the light-awakening molecules Newton ascribed the different colour-sensations. In the undulatory theory we are also confronted with the question of colour ; and here again, to inform and guide us, we have the analogy of sound. Aërial waves of different lengths, or periods, produce notes of different pitch ; and to differences of wave-length in that mysterious medium, the all-pervading ether, differences of colour are ascribed. Hooke had already discoursed of ' a very quick motion that causes light, as well as a more robust that causes heat.' Newton had ascribed the sensation of red to the shock of his grossest, and that of violet to the shock of his finest luminiferous projectiles. Defining the one, and displacing the other of these notions, the wave-theory affirms red to be produced by the largest, and violet by the smallest waves of the visible spectrum. The theory of undulation had to encounter that fierce struggle for existence which all great changes of doctrine, scientific or otherwise, have had to endure. Mighty intellects, following the mightiest of them all, were arrayed against it. But the more it was discussed the more it grew in strength and favour, until it finally supplanted its formidable rival. No competent scientific man at the present day accepts the theory of emission, or refuses to accept the theory of undulation.

Boyle and Hooke had been fruitful experimenters on those beautiful iridescences known as the 'colours of thin plates.' The rich hues of the thin-blown soap-bubble, of oil floating on water, and of the thin layer of oxide on molten lead, are familiar illustrations of these iris colours. Hooke showed that all transparent films, if only thin enough, displayed such colours; and he proved that the particular colour displayed depended upon the thickness of the film. Passing from solid and liquid films to films of air, he says: 'Take two small pieces of ground and polished looking-glass plate, each about the bigness of a shilling; take these two dry, and with your forefingers and thumbs press them very hard and close together, and you shall find that when they approach each other very near, there will appear several irises or coloured lines.' Newton, bent on knowing the exact relation between the thickness of the film and the colour it produced, varied Hooke's experiment. Taking two pieces of glass, the one plane and the other very slightly curved, and pressing both together, he obtained a film of air of gradually increasing thickness from the place of contact outwards. As he expected, he found the place of contact surrounded by a series of coloured circles, still known all over the world as ' Newton's rings.' The colours of his first circle, which immediately surrounded a black central spot, Newton called ' colours of the first order ; ' the colours of the second circle, ' colours of the second order,' and so on. With unrivalled penetration and apparent success, he applied his theory of ' fits ' to the explanation of the ' rings.' Here, however, the only immortal parts of his labours are his facts and measurements ; his theory has disappeared. It was reserved for the illustrious Thomas Young, a man of intellectual calibre resembling that of Newton himself, to prove that the rings were produced by the mutual action —in technical phrase, ' interference '—of the light-waves reflected at the two surfaces of the film of air inclosed between the plane and convex glasses. The colours of thin plates were ' residual colours '—survivals of the white light after the ravages of interference. Young soon translated the theory of ' fits ' into that of ' waves ; ' the measurements pertaining to the former being so accurate as to render them immediately available for the purposes of the latter.

It is here that Newton's researches and opinions touch the subject of this article. The colour nearest to the black spot, in the experiment above described, was a faint blue—' blue of the first order '—corresponding to the film of air when thinnest. If a solid or liquid film, of the thickness requisite to produce this colour, were broken into bits and scattered in the air, Newton inferred that the tiny fragments would display the blue colour. Tantamount to this, he considered, was the action of minute water-particles in the incipient stage of their condensation from aqueous vapour. Such particles

10

suspended in our atmosphere ought, he supposed, to
generate the serenest skies. Newton does not appear
to have bestowed much thought upon this subject; for
to produce the particular blue which he regarded as
sky-blue, thin plates with parallel surfaces would be
required. The notion that cloud-particles are hollow
spheres, or vesicles, is prevalent on the Continent, but it
never made any way among the scientific men of England. De Saussure thought that he had actually seen
the cloud-vesicles, and Faraday, as I learned from
himself, believed that he had once confirmed the observation of the illustrious Alpine traveller. During my long
acquaintance with the atmosphere of the Alps I have
often sought for these aqueous bladders, but have never
been able to find them. Clausius once published a
profound essay on the colours of the sky. The assumption of small water drops, he proved, would lead to
optical consequences entirely at variance with facts.
For a time, therefore, he closed with the idea of vesicles,
and endeavoured to deduce from them the blue of the
firmament and the morning and evening red.

It is not, however, necessary to invoke the blue of
the first order to explain the colour of the sky ; nor is
it necessary to impose upon condensing vapour the difficult, if not impossible, task of forming bladders, when
it passes into the liquid condition. Let us examine
the subject. *Eau-de-Cologne* is prepared by dissolving
aromatic gums or resins in alcohol. Dropped into
water, the scented liquid immediately produces a white
cloudiness, due to the precipitation of the substances
previously held in solution. The solid particles are,
however, comparatively gross; but by diminishing the
quantity of the dissolved gum, the precipitate may
be made to consist of extremely minute particles.
Brücke, for example, dissolved gum-mastic, in certain

proportions, in alcohol, and carefully dropping his solution into a beaker of water, kept briskly stirred, he was able to reduce the precipitate to an extremely fine state of division. The particles of mastic can by no means be imagined as forming bladders. Still, against a dark ground—black velvet, for example—the water that contains them shows a distinctly blue colour. The bluish colour of many liquids is produced in a similar manner. Thin milk is an example. Blue eyes are also said to be simply turbid media. The rocks over which glaciers pass are finely ground and pulverised by the ice, or the stony emery imbedded in it; and the river which issues from the snout of every glacier is laden with suspended matter. When such glacier water is placed in a tall glass jar, and the heavier particles are permitted to subside, the liquid column, when viewed against a dark background, has a decidedly bluish tinge. The exceptional blueness of the Lake of Geneva, which is fed with glacier water, may be due, in part, to particles small enough to remain suspended long after their larger and heavier companions have sunk to the bottom of the lake.

We need not, however, resort to water for the pro·duction of the colour. We can liberate, in air, particles of a size capable of producing a blue as deep and pure as the azure of the firmament. In fact, artificial skies may be thus generated, which prove their brotherhood with the natural sky by exhibiting all its phenomena. There are certain chemical compounds—aggregates of molecules—the constituent atoms of which are readily shaken asunder by the impact of special waves of light. Probably, if not certainly, the atoms and the waves are so related to each other, as regards vibrating period, that the wave-motion can accumulate until it becomes disruptive. A great number of substances might be

mentioned whose vapours, when mixed with air and subjected to the action of a solar or an electric beam, are
thus decomposed, the products of decomposition hanging
as liquid or solid particles in the beam which generates
them. And here I must appeal to the inner vision already
spoken of. Remembering the different sizes of the waves
of light, it is not difficult to see that our minute particles are larger with respect to some waves than to
others. In the case of water, for example, a pebble will
intercept and reflect a larger fractional part of a ripple
than of a larger wave. We have now to imagine light-
undulations of different dimensions, but all exceedingly
minute, passing through air laden with extremely small
particles. It is plain that such particles, though scattering portions of all the waves, will exert their most
conspicuous action upon the smallest ones; and that
the colour-sensation answering to the smallest waves—
in other words, the colour blue—will be predominant
in the scattered light. This harmonises perfectly with
what we observe in the firmament. The sky is blue,
but the blue is not pure. On looking at the sky
through a spectroscope, we observe all the colours of the
spectrum; blue is merely the predominant colour.
By means of our artificial skies we can take, as it were,
the firmament in our hands and examine it at our
leisure. Like the natural sky, the artificial one shows
all the colours of the spectrum, but blue in excess.
Mixing very small quantities of vapour with air, and
bringing the decomposing luminous beam into action,
we produce particles too small to shed any sensible
light, but which may, and doubtless do, exert an action
on the ultra-violet waves of the spectrum. We can
watch these particles, or rather the space they occupy,
till they grow to a size able to yield the firmamental
azure. As the particles grow larger under the continued

action of the light, the azure becomes less deep; while later on a milkiness, such as we often observe in nature, takes the place of the purer blue. Finally the particles become large enough to reflect all the light-waves, and then the suspended 'actinic cloud' diffuses white light.

It must occur to the reader that even in the absence of definite clouds there are considerable variations in the hue of the firmament. Everybody knows, moreover, that as the sky bends towards the horizon, the purer blue is impaired. To measure the intensity of the colour De Saussure invented a cyanometer, and Humboldt has given us a mathematical formula to express the diminution of the blue, in arcs drawn east and west from the zenith downwards. This diminution is a natural consequence of the predominance of coarser particles in the lower regions of the atmosphere. Were the particles which produce the purer celestial vault all swept away, we should, unless helped by what has been called 'cosmic dust,' look into the blackness of celestial space. And were the whole atmosphere abolished along with its suspended matter, we should have the 'blackness' spangled with steady stars; for the twinkling of the stars is caused by our atmosphere. Now, the higher we ascend, the more do we leave behind us the particles which scatter the light; the nearer, in fact, do we approach to that vision of celestial space mentioned a moment ago. Viewed, therefore, from the loftiest Alpine summits, the firmamental blue is darker than it is ever observed to be from the plains.

It is thus shown that by the scattering action of minute particles the blue of the sky can be produced; but there is yet more to be said upon the subject. Let the natural sky be looked at on a fine day through a piece of transparent Iceland spar cut into the form known as a Nicol prism. It may be well to begin by

looking through the prism at a snow slope, or a white
wall. Turning the prism round its axis, the light coming from these objects does not undergo any sensible
change. But when the prism is directed towards the
sky the great probability is that, on turning it, variations
in the amount of light reaching the eye will be observed.
Testing various portions of the sky with due diligence,
we at length discover one particular direction where the
difference of illumination becomes a maximum. Here
the Nicol, in one position, seems to offer no impediment
to the passage of the sky light; while, when turned
through an arc of ninety degrees from this position, the
light is almost entirely quenched. We soon discern that
the particular line of vision in which this maximum
difference is observed is perpendicular to the direction
of the solar rays. The Nicol acts thus upon sky light
because that light is polarised, while the light from the
white wall or the white snow, being unpolarised, is not
affected by the rotation of the prism.

In the case of our manufactured sky not only is the
azure of the firmament reproduced, but these phenomena
of polarisation are observed even more perfectly than
in the natural sky. When the air-space from which
our best artificial azure is emitted is examined with the
Nicol prism, the blue light is found to be completely
polarised at right angles to the illuminating beam.
The artifical sky may, in fact, be employed as a second
Nicol, between which and a prism held in the hand
many of the beautiful chromatic phenomena observed
in an ordinary polariscope may be reproduced.

Let us now complete our thesis by following the
larger light-waves, which have been able to pass among
the aërial particles with comparatively little fractional
loss. Without going beyond inferential considerations,
we can state what must occur. The action of the

particles upon the solar light increases with the atmospheric distances traversed by the sun's rays. The lower the sun, therefore, the greater the action. The shorter waves of the spectrum being more and more withdrawn, the tendency is to give the longer waves an enchanced predominance in the transmitted light. The tendency, in other words, of this light, as the rays traverse ever-increasing distances, is more and more towards red. This, I say, might be stated as an inference, but it is borne out in the most impressive manner by facts. When the Alpine sun is setting, or, better still, some time after he has set, leaving the limbs and shoulders of the mountains in shadow, while their snowy crests are bathed by the retreating light, the snow glows with a beauty and solemnity hardly equalled by any other natural phenomenon. So, also, when first illumined by the rays of the unrisen sun, the mountain heads, under favourable atmospheric conditions, shine like rubies. And all this splendour is evoked by the simple mechanism of minute particles, themselves without colour, suspended in the air. Those who referred the extraordinary succession of atmospheric glows, witnessed some years ago, to a vast and violent discharge of volcanic ashes, were dealing with 'a true cause.' The fine floating residue of such ashes would, undoubtedly, be able to produce the effects ascribed to it. Still, the mechanism necessary to produce the morning and the evening red, though of variable efficiency, is always present in the atmosphere. I have seen displays, equal in magnificence to the finest of those above referred to, when there was no special volcanic outburst to which they could be referred. It was the long-continued repetition of the glows which rendered the volcanic theory highly probable.

VI.

VOYAGE TO ALGERIA TO OBSERVE THE ECLIPSE.

1870.

THE opening of the Eclipse Expedition was not pro-
pitious. Portsmouth, on Monday, December 5,
1870, was swathed by fog, which was intensified by smoke,
and traversed by a drizzle of fine rain. At six P.M. I
was on board the ' Urgent.' On Tuesday morning the
weather was too thick to permit of the ship's being
swung and her compasses calibrated. The Admiral of
the port, a man of very noble presence, came on board.
Under his stimulus the energy which the weather had
damped appeared to become more active, and soon after
his departure we steamed down to Spithead. Here the
fog had so far lightened as to enable the officers to swing
the ship.

At three P.M. on Tuesday, December 6, we got away,
gliding successively past Whitecliff Bay, Bembridge,
Sandown, Shanklin, Ventnor, and St. Catherine's Lighthouse. On Wednesday morning we sighted the Isle of
Ushant, on the French side of the Channel. The
northern end of the island has been fretted by the
waves into detached tower-like masses of rock of very
remarkable appearance. In the Channel the sea was
green, and opposite Ushant it was a brighter green. On
Wednesday evening we committed ourselves to the Bay
of Biscay. The roll of the Atlantic was full, but not

violent. There had been scarcely a gleam of sunshine throughout the day, but the cloud-forms were fine, and their apparent solidity impressive. On Thursday morning the green of the sea was displaced by a deep indigo blue. The whole of Thursday we steamed across the bay. We had little blue sky, but the clouds were again grand and varied—cirrus, stratus, cumulus, and nimbus, we had them all. Dusky hair-like trails were sometimes dropped from the distant clouds to the sea. These were falling showers, and they sometimes occupied the whole horizon, while we steamed across the rainless circle which was thus surrounded. Sometimes we plunged into the rain, and once or twice, by slightly changing our course, avoided a heavy shower. From time to time perfect rainbows spanned the heavens from side to side. At times a bow would appear in fragments, showing the keystone of the arch midway in air, and its two buttresses on the horizon. In all cases the light of the bow could be quenched by a Nicol's prism, with its long diagonal tangent to the arc. Sometimes gleaming patches of the firmament were seen amid the clouds. When viewed in the proper direction, the gleam could be quenched by a Nicol's prism, a dark aperture being thus opened into stellar space.

At sunset on Thursday the denser clouds were fiercely fringed, while through the lighter ones seemed to issue the glow of a conflagration. On Friday morning we sighted Cape Finisterre—the extreme end of the arc which sweeps from Ushant round the Bay of Biscay. Calm spaces of blue, in which floated quietly scraps of cumuli, were behind us, but in front of us was a horizon of portentous darkness. It continued thus threatening throughout the day. Towards evening the wind strengthened to a gale, and at dinner it was difficult to preserve the plates and dishes from destruction. Our thinned company hinted that the rolling had other consequences. It was very wild when we went to bed. I slumbered and slept, but after some time was rendered anxiously conscious that my body had become a kind of projectile, with the ship's side for a target. I gripped the edge of my berth to save myself from being thrown out. Outside, I could hear somebody say that he had been thrown from his berth, and sent spinning to the other side of the saloon. The screw laboured violently amid the lurching; it incessantly quitted the water, and, twirling in the air, rattled against its bearings, causing the ship to shudder from stem to stern. At times the waves struck us, not with the soft impact which might be expected from a liquid, but with the sudden solid shock of battering-rams. 'No man knows the force of water,' said one of the officers, 'until he has experienced a storm at sea.' These blows followed each other at quicker intervals, the screw rattling after each of them, until, finally, the delivery of a heavier stroke than ordinary seemed to reduce the saloon to chaos. Furniture crashed, glasses rang, and alarmed enquiries immediately followed. Amid the noises I heard one note of forced laughter; it sounded very ghastly. Men tramped through the saloon, and busy voices were heard aft, as if something there had gone wrong.

I rose, and not without difficulty got into my clothes In the after-cabin, under the superintendence of the able and energetic navigating lieutenant, Mr. Brown, a group of blue-jackets were working at the tiller-ropes. These had become loose, and the helm refused to answer the wheel. High moral lessons might be gained on shipboard, by observing what stead-fast adherence to an object can accomplish, and what large effects are heaped up by the addition of infinitesimals. The tiller-rope, as the blue-jackets strained in concert, seemed hardly to move; still it did move a little, until finally, by timing the pull to the lurching of the ship, the mastery of the rudder was obtained. I had previously gone on deck. Round the saloon-door were a few members of the eclipse party, who seemed in no mood for scientific observation. Nor did I; but I wished to see the storm. I climbed the steps to the poop, exchanged a word with Captain Toynbee, the only member of the party to be seen on the poop, and by his direction made towards a cleat not far from the wheel.[1] Round it I coiled my arms. With the exception of the men at the wheel, who stood as silent as corpses, I was alone.

I had seen grandeur elsewhere, but this was a new form of grandeur to me. The 'Urgent' is long and narrow, and during our expedition she lacked the steadying influence of sufficient ballast. She was for a time practically rudderless, and lay in the trough of the sea. I could see the long ridges, with some hundreds of feet between their crests, rolling upon the ship perfectly parallel to her sides. As they approached, they so grew upon the eye as to render the expression 'mountains high' intelligible. At all events, there was no mistaking their mechanical might, as they took the ship upon their shoulders, and swung her like a pendulum. The deck sloped sometimes at an angle which I estimated at over forty-five degrees; wanting my previous Alpine practice, I should have felt less confidence in my grip of the cleat. Here and there the long rollers were tossed by interference into heaps of greater height. The wind caught their crests, and scattered them over the sea, the whole surface of which was seething white. The aspect

[1] The cleat is a T-shaped mass of metal employed for the fastening of ropes.

of the clouds was a fit accompaniment to the fury of the ocean. The moon was almost full—at times concealed, at times revealed, as the scud flew wildly over it. These things appealed to the eye, while the ear was filled by the groaning of the screw and the whistle and boom of the storm.

Nor was the outward agitation the only object of interest to me. I was at once subject and object to myself, and watched with intense interest the workings of my own mind. The 'Urgent' is an elderly ship. She had been built, I was told, by a contracting firm for some foreign Government, and had been diverted from her first purpose when converted into a troop-ship. She had been for some time out of work, and I had heard that one of her boilers, at least, needed repair. Our scanty but excellent crew, moreover, did not belong to the 'Urgent,' but had been gathered from other ships. Our three lieutenants were also volunteers. All this passed swiftly through my mind as the steamer shook under the blows of the waves, and I thought that probably no one on board could say how much of this thumping and straining the 'Urgent' would be able to bear. This uncertainty caused me to look steadily at the worst, and I tried to strengthen myself in the face of it.

But at length the helm laid hold of the water, and the ship was got gradually round to face the waves. The rolling diminished, a certain amount of pitching taking its place. Our speed had fallen from eleven knots to two. I went again to bed. After a space of calm, when we seemed crossing the vortex of a storm, heavy tossing recommenced. I was afraid to allow myself to fall asleep, as my berth was high, and to be pitched out of it might be attended with bruises, if not with fractures. From Friday at noon to Saturday at noon we accomplished sixty-six miles, or an average of less than three miles an hour. I overheard the sailors talking about this storm. The 'Urgent,' according to those that knew her, had never previously experienced anything like it.[1]

All through Saturday the wind, though somewhat sobered, blew dead against us. The atmospheric effects were exceedingly fine. The cumuli resembled mountains in shape, and their peaked summits shone as white as Alpine snows. At one place this resemblance was greatly strengthened by a vast area of cloud, uniformly illuminated, and lying like a *névé* below the peaks. From it fell a kind of cloud-river strikingly like a glacier. The horizon at sunset was remarkable—spaces of brilliant green between clouds of fiery red. Rainbows had been frequent throughout the day, and at night a perfectly continuous lunar bow spanned the heavens from side to side. Its colours were feeble; but, contrasted with the black ground against which it rested, its luminousness was extraordinary.

Sunday morning found us opposite to Lisbon, and at midnight we rounded Cape St. Vincent, where the lurching seemed disposed to recommence. Through the kindness of Lieutenant Walton, a cot had been slung for me. It hung between a tiller-wheel and a flue, and at one A.M. I was roused by the banging of the cot against its boundaries. But the wind was now behind us, and we went along at a speed of eleven knots. We felt certain of reaching Cadiz by three. But a new lighthouse came in sight, which some affirmed to be Cadiz Lighthouse, while the surrounding houses were declared to be those of Cadiz itself. Out of deference to

---

[1] There is, it will be seen, a fair agreement between these impressions and those so vigorously described by a scientific correspondent of the 'Times.'

these statements, the navigating lieutenant changed his course, and steered for the place. A pilot came on board, and he informed us that we were before the mouth of the Guadalquivir, and that the lighthouse was that of Cipiòna. Cadiz was still some eighteen miles distant.

We steered towards the city, hoping to get into the harbour before dark. But the pilot who would have guided us had been snapped up by another vessel, and we did not get in. We beat about during the night, and in the morning found ourselves about fifteen miles from Cadiz. The sun rose behind the city, and we steered straight into the light. The three-towered cathedral stood in the midst, round which swarmed apparently a multitude of chimney-stacks. A nearer approach showed the chimneys to be small turrets. A pilot was taken on board; for there is a dangerous shoal in the harbour. The appearance of the town as the sun shone upon its white and lofty walls was singularly beautiful. We cast anchor; some officials arrived and demanded a clean bill of health. We had none. They would have nothing to do with us; so the yellow quarantine flag was hoisted, and we waited for permission to land the Cadiz party. After some hours' delay the English consul and vice-consul came on board, and with them a Spanish officer ablaze with gold lace and decorations. Under slight pressure the requisite permission had been granted. We landed our party, and in the afternoon weighed anchor. Thanks to the kindness of our excellent paymaster, I was here transferred to a more roomy berth.

Cadiz soon sank beneath the sea, and we sighted in succession Cape Trafalgar, Tarifa, and the revolving light of Ceuta. The water was very calm, and the moon rose in a quiet heaven. She swung with her convex surface downwards, the common boundary between light and shadow being almost horizontal. A pillar of reflected light shimmered up to us from the slightly rippled sea. I had previously noticed the phosphorescence of the water, but to night it was stronger than usual, especially among the foam at the bows. A bucket let down into the sea brought up a number of the little sparkling organisms which caused the phosphorescence. I caught some of them in my hand. And here an appearance was observed which was new to most of us, and strikingly beautiful to all. Standing at the bow and looking forwards, at a distance of forty or fifty yards from the ship, a number of luminous streamers were seen rushing towards us. On nearing the vessel they rapidly turned, like a comet round its perihelion, placed themselves side by side, and, in parallel trails of light, kept up with the ship. One of them placed itself right in front of the bow as a pioneer. These comets of the sea were joined at intervals by others. Sometimes as many as six at a time would rush at us, bend with extraordinary rapidity round a sharp curve, and afterwards keep us company. I leaned over the bow, and scanned the streamers closely. The frontal portion of each of them revealed the outline of a porpoise. The rush of the creatures through the water had started the phosphorescence, every spark of which was converted by the motion of the retina into a line of light. Each porpoise was thus wrapped in a luminous sheath. The phosphorescence did not cease at the creature's tail, but was carried many porpoise-lengths behind it.

To our right we had the African hills, illuminated by the moon. Gibraltar Rock at length became visible, but the town remained long hidden by a belt of haze, through which at length the brighter lamps struggled. It was like the gradual resolution of a nebula into stars. As the intervening depth became gradually less, the mist vanished more and more, and finally all the lamps shone through it   They formed a bright foil to the sombre mass of rock above them.   The sea was so calm and the scene so lovely that Mr. Huggins and myself stayed on deck till near midnight, when the ship was moored.   During our walking to and fro a striking enlargement of the disk of Jupiter was observed, whenever the heated air of the funnel came between us and the planet.   On passing away from the heated air, the flat dim disk would immediately shrink to a luminous point.   The effect was one of visual persistence.   The retinal image of the planet was set quivering in all azimuths by the streams of heated air, describing in quick succession minute lines of light, which summed themselves to a disk of sensible area.

At six o'clock next morning, the gun at the Signal Station on the summit of the rock, boomed.   At eight the band on board the 'Trafalgar' training-ship, which was in the harbour, struck up the national anthem; and immediately afterwards a crowd of mite-like cadets swarmed up the rigging.   After the removal of the apparatus belonging to the Gibraltar party we went on shore.   Winter was in England when we left, but here we had the warmth of summer.   The vegetation was luxuriant—palm-trees, cactuses, and aloes, all ablaze with scarlet flowers.   A visit to the Governor was proposed, as an act of necessary courtesy, and I accompanied Admiral Ommaney and Mr. Huggins to 'the Convent,' or Government House.   We sent in our cards, waited for a time, and were then conducted by an orderly to his Excellency.   He is a fine old man, over six feet high, and of frank military bearing.   He received us and conversed with us in a very genial manner.   He took us to see his garden, his palms, his shaded promenades, and his orange-trees loaded with fruit, in all of which he took manifest delight. Evidently 'the hero of Kars' had fallen upon quarters after his own heart. He appeared full of good nature, and engaged us on the spot to dine with him that day.

We sought the town-major for a pass to visit the lines. While awaiting his arrival I purchased a stock of white glass bottles, with a view to experiments on the colour of the sea. Mr. Huggins and myself, who wished to see the rock, were taken by Captain Salmond to the library, where a model of Gibraltar is kept, and where we had a useful preliminary lesson. At the library we met Colonel Maberly, a courteous and kindly man, who gave us good advice regarding our excursion. He sent an orderly with us to the entrance of the lines. The orderly handed us over to an intelligent Irishman, who was directed to show us everything that we desired to see, and to hide nothing from us. We took the 'upper line,' traversed the galleries hewn through the limestone; looked through the embrasures, which opened like doors in the precipice, towards the hills of Spain; reached St. George's hall, and went still higher, emerging on the summit of one of the noblest cliffs I have ever seen.

Beyond were the Spanish lines, marked by a line of white sentry-boxes; nearer were the English lines, less conspicuously indicated; and between both was the neutral ground. Behind the Spanish lines rose the conical hill called the Queen of Spain's Chair. The general aspect of the mainland from the rock is bold and rugged. Doubling back from the galleries, we struck upwards towards the crest, reached the Signal Station, where we indulged in 'shandy-gaff' and bread and cheese. Thence to O'Hara's Tower, the highest point of the rock. It was built by a former Governor, who, forgetful of the laws of terrestrial curvature, thought he

11

might look from the tower into the port of Cadiz.   The
tower is riven, and it may be climbed along the edges
of the crack.   We got to the top of it; thence descended the curious Mediterranean Stair—a zigzag,
mostly of steps down a steeply falling slope, amid
palmetto brush, aloes, and prickly pear.

Passing over the Windmill Hill, we were joined at
the ' Governor's Cottage ' by a car, and drove afterwards
to the lighthouse at Europa Point.   The tower was
built, I believe, by Queen Adelaide, and it contains a
fine dioptric apparatus of the first order, constructed by
Messrs. Chance, of Birmingham.   At the appointed
hour we were at the Convent.   During dinner the same
genial traits which appeared in the morning were still
more conspicuous.   The freshness of the Governor's
nature showed itself best when he spoke of his old antagonist in arms, Mouravieff.   Chivalry in war is con-
sistent with its stern prosecution.   These two men
were chivalrous, and after striking the last blow became
friends for ever.   Our kind and courteous reception at
Gibraltar is a thing to be remembered with pleasure.

On December 15 we committed ourselves to the
Mediterranean.   The views of Gibraltar with which we
are most acquainted represent it as a huge ridge ; but
its aspect, end on, both from the Spanish lines and from
the other side, is truly noble.   There is a sloping bank
of sand at the back of the rock, which I was disposed
to regard simply as the *débris* of the limestone.   I
wished to let myself down upon it, but had not the
time.   My friend Mr. Busk, however, assures me that
it is silica, and that the same sand constitutes the adjacent neutral ground.   There are theories afloat as to
its having been blown from Sahara.   The Mediterranean
throughout this first day, and indeed throughout the
entire voyage to Oran, was of a less deep blue than the

Atlantic. Possibly the quantity of organisms may have modified the colour. At night the phosphorescence was startling, breaking suddenly out along the crests of the waves formed by the port and starboard bows. Its strength was not uniform. Having flashed brilliantly for a time, it would in part subside, and afterwards regain its vigour. Several large phosphorescent masses of weird appearance also floated past.

On the morning of the 16th we sighted the fort and lighthouse of Marsa el Kibir, and beyond them the white walls of Oran lying in the bight of a bay, sheltered by dominant hills. The sun was shining brightly ; during our whole voyage we had not had so fine a day. The wisdom which had led us to choose Oran as our place of observation seemed demonstrated. A rather excitable pilot came on board, and he guided us in behind the Mole, which had suffered much damage the previous year from an unexplained outburst of waves from the Mediterranean. Both port and bow anchors were cast in deep water. With three huge hawsers the ship's stern was made fast to three gun-pillars fixed in the Mole ; and here for a time the ' Urgent' rested from her labours.

M. Janssen, who had rendered his name celebrated by his observations of the eclipse in India in 1868, when he showed the solar flames to be eruptions of incandescent hydrogen, was already encamped in the open country about eight miles from Oran. On December 2 he had quitted Paris in a balloon, with a strong young sailor as his assistant, had descended near the mouth of the Loire, seen M. Gambetta, and received from him encouragement and aid. On the day of our arrival his encampment was visited by Mr. Huggins, and the kind and courteous Engineer of the Port drove me subsequently, in his own phaeton, to the place. It bore the best repute as regards freedom from haze and fog, and commanded an open outlook; but it was inconvenient for us on account of its distance from the ship. The place next in repute was the railway station, between two and three miles distant from the Mole. It was inspected, but, being enclosed, was abandoned for an eminence in an adjacent garden, the property of Mr. Hinshelwood, a Scotchman who had settled some years previously as an Esparto merchant in Oran.[1] He, in the most liberal manner, placed his ground at the disposition of the party. Here the tents were pitched, on the Saturday, by Captain Salmond and his intelligent corps of sappers, the instruments being erected on the Monday under cover of the tents.

Close to the railway station runs a new loopholed wall of defence, through which the highway passes into the open country. Standing on the highway, and looking southwards, about twenty yards to the right is a small bastionet, intended to carry a gun or two. Its roof I thought would form an admirable basis for my telescope, while the view of the surrounding country was unimpeded in all directions. The authorities kindly allowed me the use of this bastionet. Two men, one a blue-jacket named Elliot, and the other a marine named Hill, were placed at my disposal by Lieutenant Walton; and, thus aided, on Monday morning I mounted my telescope. The instrument was new to me, and some hours of discipline were spent in mastering all the details of its manipulation.

Mr. Huggins joined me, and we visited together the Arab quarter of Oran. The flat-roofed houses appeared very clean and white. The street was filled with loiterers, and the thresholds were occupied by pictur-

[1] Esparto is a kind of grass now much used in the manufacture of paper.

csque groups. Some of the men were very fine. We saw many straight, manly fellows who must have been six feet four in height. They passed us with perfect indifference, evincing no anger, suspicion, or curiosity, hardly caring in fact to glance at us as we passed. In one instance only during my stay at Oran was I spoken to by an Arab. He was a tall, good-humoured fellow, who came smiling up to me, and muttered something about ' les Anglais.' The mixed population of Oran is picturesque in the highest degree : the Jews, rich and poor, varying in their costumes as their wealth varies ; the Arabs more picturesque still, and of all shades of complexion—the negroes, the Spaniards, the French, all grouped together, each race preserving its own individuality, formed a picture intensely interesting to me.

On Tuesday, the 20th, I was early at the bastionet. The night had been very squally. The sergeant of the sappers had taken charge of our key, and on Tuesday morning Elliot went for it. He brought back the intelligence that the tents had been blown down, and the instruments overturned. Among these was a large and valuable equatorial from the Royal Observatory, Greenwich. It seemed hardly possible that this instrument, with its wheels and verniers and delicate adjustments, could have escaped uninjured from such a fall. This, however, was the case ; and during the day all the overturned instruments were restored to their places, and found to be in practical working order. This and the following day were devoted to incessant schooling. I had come out as a general stargazer, and not with the intention of devoting myself to the observation of any particular phenomenon. I wished to see the whole— the first contact, the advance of the moon, the successive swallowing up of the solar spots, the breaking of the last line of crescent by the lunar mountains into

Bailey's beads, the advance of the shadow through the air, the appearance of the corona and prominences at the moment of totality, the radiant streamers of the corona, the internal structure of the flames, a glance through a polariscope, a sweep round the landscape with the naked eye, the reappearance of the solar limb through Bailey's beads, and, finally, the retreat of the lunar shadow through the air.

I was provided with a telescope of admirable definition, mounted, adjusted, packed, and most liberally placed at my disposal by Mr. Warren De La Rue. The telescope grasped the whole of the sun, and a considerable portion of the space surrounding it. But it would not take in the extreme limits of the corona. For this I had lashed on to the large telescope a light but powerful instrument, constructed by Ross, and lent to me by Mr. Huggins. I was also furnished with an excellent binocular by Mr. Dallmeyer. In fact, no man could have been more efficiently supported. It required a strict parcelling out of the interval of totality to embrace in it the entire series of observations. These, while the sun remained visible, were to be made with an unsilvered diagonal eye-piece, which reflected but a small fraction of the sun's light, this fraction being still further toned down by a dark glass. At the moment of totality the dark glass was to be removed, and a silver reflector pushed in, so as to get the maximum of light from the corona and prominences. The time of totality was distributed as follows:

1. Observe approach of shadow through the air: totality.
2. Telescope . . . . . 30 seconds.
3. Finder . . . . . 30 seconds.
4. Double image prism . . 15 seconds.
5. Naked eye . . . . 10 seconds.
6. Finder or binocular . . 20 seconds.
7. Telescope . . . 20 seconds.
8. Observe retreat of shadow.

In our rehearsals Elliot stood beside me, watch in hand, and furnished with a lantern. He called out at the end of each interval, while I moved from telescope to finder, from finder to polariscope, from polariscope to naked eye, from naked eye back to finder, from finder to telescope, abandoning the instrument finally to observe the retreating shadow. All this we went over twenty times, while looking at the actual sun, and keeping him in the middle of the field. It was my object to render the repetition of the lesson so mechanical as to leave no room for flurry, forgetfulness, or excitement. Volition was not to be called upon, nor judgment exercised, but a well-beaten path of routine was to be followed. Had the opportunity occurred, I think the programme would have been strictly carried out.

But the opportunity did not occur. For several days the weather had been ill-natured. We had wind so strong as to render the hawsers at the stern of the ' Urgent ' as rigid as iron, and to destroy the navigating lieutenant's sleep. We had clouds, a thunder-storm, and some rain. Still the hope was held out that the atmosphere would cleanse itself, and if it did we were promised air of extraordinary limpidity. Early on the 22nd we were all at our posts. Spaces of blue in the early morning gave us some encouragement, but all depended on the relation of these spaces to the surrounding clouds. Which of them were to grow as the day advanced ? The wind was high, and to secure the steadiness of my instrument I was forced to retreat behind a projection of the bastionet, place stones upon its stand, and, further, to avail myself of the shelter of a sail. My practised men fastened the sail at the top, and loaded it with boulders at the bottom. It was tried severely, but it stood firm.

The clouds and blue spaces fought for a time with varying success. The sun was hidden and revealed at intervals, hope oscillating in synchronism with the changes of the sky. At the moment of first contact a dense cloud intervened; but a minute or two afterwards the cloud had passed, and the encroachment of the black body of the moon was evident upon the solar disk. The moon marched onward, and I saw it at frequent intervals; a large group of spots were approached and swallowed up. Subsequently I caught sight of the lunar limb as it cut through the middle of a large spot. The spot was not to be distinguished from the moon, but rose like a mountain above it. The clouds, when thin, could be seen as grey scud drifting across the black surface of the moon; but they thickened more and more, and made the intervals of clearness scantier. During these moments I watched with an interest bordering upon fascination the march of the silver sickle of the sun across the field of the telescope. It was so sharp and so beautiful. No trace of the lunar limb could be observed beyond the sun's boundary. Here, indeed, it could only be relieved by the corona, which was utterly cut off by the dark glass. The blackness of the moon beyond the sun was, in fact, confounded with the blackness of space.

Beside me was Elliot with the watch and lantern, while Lieutenant Archer, of the Royal Engineers, had the kindness to take charge of my note-book. I mentioned, and he wrote rapidly down, such things as seemed worthy of remembrance. Thus my hands and mind were entirely free; but it was all to no purpose. A patch of sunlight fell and rested upon the landscape some miles away. It was the only illuminated spot within view. But to the north-west there was still a space of blue which might reach us in time. Within seven minutes of totality another space towards the zenith became very dark. The atmosphere was, as it were, on the brink of a precipice, being charged with humidity, which required but a slight chill to bring it down in clouds. This was furnished by the withdrawal of the solar beams: the clouds did come down, covering up the space of blue on which our hopes had so long rested. I abandoned the telescope and walked to and fro in despair. As the moment of totality approached, the descent towards darkness was as obvious as a falling stone. I looked towards a distant ridge, where the darkness would first appear. At the moment a fan of beams, issuing from the hidden sun, was spread out over the southern heavens. These beams are bars of alternate light and shade, produced in illuminated haze by the shadows of floating cloudlets of varying density. The beams are practically parallel, but by an effect of perspective they appear divergent, having the sun, in fact, for their point of convergence. The darkness took possession of the ridge referred to, lowered upon M. Janssen's observatory, passed over the southern heavens, blotting out the beams as if a sponge had been drawn across them. It then took successive possession of three spaces of blue sky in the south-eastern atmosphere. I again looked towards the ridge. A glimmer as of day-dawn was behind it, and immediately afterwards the fan of beams, which had been for more than two minutes absent, revived. The eclipse of 1870 had ended, and, as far as the corona and flames were concerned, we had been defeated.

Even in the heart of the eclipse the darkness was by no means perfect. Small print could be read. In fact, the clouds which rendered the day a dark one, by scattering light into the shadow, rendered the darkness less intense than it would have been had the atmosphere been without cloud. In the more open spaces I sought for stars, but could find none. There was a lull in the wind before and after totality, but during the totality the wind was strong. I waited for some time on the bastionet, hoping to get a glimpse of the moon on the opposite border of the sun, but in vain. The clouds continued, and some rain fell. The day brightened somewhat afterwards, and, having packed all up, in the sober twilight Mr. Crookes and myself climbed the heights above the fort of Vera Cruz. From this eminence we had a very noble view over the Mediterranean and the flanking African hills. The sunset was remarkable, and the whole outlook exceedingly fine.

The able and well-instructed medical officer of the 'Urgent,' Mr. Goodman, observed the following temperatures during the progress of the eclipse :

Hour		Deg.	Hour		Deg.
11.45	. .	56	12.43	. .	51
11.55	. .	55	1.5	. .	52
12.10	. .	54	1.27	. .	53
12.37	. .	53	1.44	. .	56
12.39	. .	52	2.10	. .	57

The minimum temperature occurred some minutes after totality, when a slight rain fell.

The wind was so strong on the 23rd that Captain Henderson would not venture out. Guided by Mr. Goodman, I visited a cave in a remarkable stratum of shell-breccia, and, thanks to my guide, secured specimens. Mr. Busk informs me that a precisely similar breccia is found at Gibraltar, at approximately the same level. During the afternoon, Admiral Ommaney and myself drove to the fort of Marsa el Kibir. The fortification is of ancient origin, the Moorish arches being still there in decay, but the fort is now very strong. About four or five hundred fine-looking dragoons were

looking after their horses, waiting for a lull to enable
them to embark for France. One of their officers was
wandering in a very solitary fashion over the fort. We
had some conversation with him. He had been at
Sedan, had been taken prisoner, but had effected his
escape. He shook his head when we spoke of the termination of the war, and predicted its long continuance.
There was bitterness in his tone as he spoke of the
charges of treason so lightly levelled against French
commanders. The green waves raved round the promontory on which the fort stands, smiting the rocks,
breaking into foam, and jumping, after impact, to a
height of a hundred feet and more into the air. As we
returned our vehicle broke down through the loss of a
wheel. The Admiral went on board, while I remained
long watching the agitated sea. The little horses of
Oran well merit a passing word. Their speed and en-
durance, both of which are heavily drawn upon by their
drivers, are extraordinary.

The wind sinking, we lifted anchor on the 24th.
For some hours we went pleasantly along; but during
the afternoon the storm revived, and it blew heavily
against us all the night. When we came opposite the
Bay of Almeria, on the 25th, the captain turned the
ship, and steered into the bay, where, under the shadow
of the Sierra Nevada, we passed Christmas night in
peace. Next morning 'a rose of dawn' rested on the
snows of the adjacent mountains, while a purple haze
was spread over the lower hills. I had no notion that
Spain possessed so fine a range of mountains as the
Sierra Nevada. The height is considerable, but the
form also is such as to get the maximum of grandeur
out of the height. We weighed anchor at eight A.M.,
passing for a time through shoal water, the bottom
having been evidently stirred up. The adjacent land

seemed eroded in a remarkable manner. It has its floods, which excavate these valleys and ravines, and leave those singular ridges behind. Towards evening I climbed the mainmast, and, standing on the cross-trees, saw the sun set amid a blaze of fiery clouds. The wind was strong and bitterly cold, and I was glad to slide back to the deck along a rope, which stretched from the mast-head to the ship's side. That night we cast anchor beside the Mole of Gibraltar.

On the morning of the 27th, in company with two friends, I drove to the Spanish lines, with the view of seeing the rock from that side. It is an exceedingly noble mass. The Peninsular and Oriental mail-boat had been signalled and had come. Heavy duties called me homeward, and by transferring myself from the 'Urgent' to the mail-steamer I should gain three days. I hired a boat, rowed to the steamer, learned that she was to start at one, and returned with all speed to the 'Urgent.' Making known to Captain Henderson my wish to get away, he expressed doubts as to the possibility of reaching the mail-steamer in time. With his accustomed kindness, he however placed a boat at my disposal. Four hardy fellows and one of the ship's officers jumped into it; my luggage, hastily thrown together, was tumbled in, and we were immediately on our way. We had nearly four miles to row in about twenty minutes; but we hoped the mail-boat might not be punctual. For a time we watched her anxiously; there was no motion; we came nearer, but the flags were not yet hauled in. The men put forth all their strength, animated by the exhortations of the officer at the helm. The roughness of the sea rendered their efforts to some extent nugatory: still we were rapidly approaching the steamer. At length she moved, punctual almost to the minute, at first slowly, but soon with quickened pace.

We turned to the left, so as to cut across her bows. Five minutes' pull would have brought us up to her. The officer waved his cap and I my hat. ' If they could only see us, they might back to us in a moment.' But they did not see us, or if they did, they paid us no attention. I returned to the ' Urgent,' discomfited, but grateful to the fine fellows who had wrought so hard to carry out my wishes.

Glad of the quiet, in the sober afternoon I took a walk towards Europa Point. The sky darkened and heavy squalls passed at intervals. Private theatricals were at the Convent, and the kind and courteous Governor had sent cards to the eclipse party. I failed in my duty in not going. St. Michael's Cave is said to rival, if it does not outrival, the Mammoth Cave of Kentucky. On the 28th Mr. Crookes, Mr. Carpenter, and myself, guided by a military policeman who understood his work, explored the cavern. The mouth is about 1,100 feet above the sea. We zigzagged up to it, and first were led into an aperture in the rock, at some height above the true entrance of the cave. In this upper cavern we saw some tall and beautiful stalactite pillars.

The water drips from the roof charged with bicarbonate of lime. Exposed to the air, the carbonic acid partially escapes, and the simple carbonate of lime, which is hardly at all soluble in water, deposits itself as a solid, forming stalactites and stalagmites. Even the exposure of chalk or limestone water to the open air partially softens it. A specimen of the Redbourne water exposed by Professors Graham, Miller, and Hofmann, in a shallow basin, fell from eighteen degrees to nine degrees of hardness. The softening process of Clark is virtually a hastening of the natural process. Here, however, instead of being permitted to evaporate, half the carbonic acid is appropriated by lime, the half thus taken up, as well as the remaining half, being precipitated. The solid precipitate is permitted to sink, and the clear supernatant liquid is limpid soft water.

We returned to the real mouth of St. Michael's Cave, which is entered by a wicket. The floor was somewhat muddy, and the roof and walls were wet. We soon found ourselves in the midst of a natural temple, where tall columns sprang complete from floor to roof, while incipient columns were growing to meet each other, upwards and downwards. The water which trickles from the stalactite, after having in part yielded up its carbonate of lime, falls upon the floor vertically underneath, and there builds the stalagmite. Consequently, the pillars grow from above and below simultaneously, along the same vertical. It is easy to distinguish the stalagmitic from the stalactitic portion of the pillars. The former is always divided into short segments by protuberant rings, as if deposited periodically, while the latter presents a uniform surface. In some cases the points of inverted cones of stalactite rested on the centres of pillars of stalagmite. The process of solidification and the consequent architecture were alike beautiful.

We followed our guide through various branches and arms of the cave, climbed and descended steps, halted at the edges of dark shafts and apertures, and squeezed ourselves through narrow passages. From time to time we halted, while Mr. Crookes illuminated with ignited magnesium wire, the roof, columns, dependent spears, and graceful drapery of the stalactites. Once, coming to a magnificent cluster of icicle-like spears, we helped ourselves to specimens. There was some difficulty in detaching the more delicate ones, their fragility was so great. A consciousness of vandalism, which smote me at the time, haunts me still; for, though our requisitions were moderate, this beauty ought not to be at all invaded. Pendent from the roof, in their natural habitat, nothing can exceed their delicate beauty; they *live*, as it were, surrounded by organic connections. In London they are curious, but not beautiful. Of gathered shells Emerson writes:

> I wiped away the weeds and foam,
> And brought my sea-born treasures home:
> But the poor, unsightly, noisome things
> Had left their beauty on the shore,
> With the sun, and the sand, and the wild uproar.

The promontory of Gibraltar is so burrowed with caverns that it has been called the Hill of Caves. They are apparently related to the geologic disturbances which the rock has undergone. The earliest of these is the tilting of the once horizontal strata. Suppose a force of torsion to act upon the promontory at its southern extremity near Europa Point, and suppose the rock to be of a partially yielding character; such a force would twist the strata into screw-surfaces, the greatest amount of twisting being endured near the point of application of the force. Such a twisting the rock appears to have suffered; but instead of the twist fading gradually and uniformly off, in passing from south to north, the want of uniformity in the material has produced lines of dislocation where there are abrupt changes in the amount of twist. Thus, at the northern end of the rock the dip to the west is nineteen degrees; in the Middle Hill, it is thirty-eight degrees; in the centre of·the South hill, or Sugar Loaf, it is fifty-seven degrees. At the southern extremity of the Sugar Loaf the strata are vertical, while farther to the south they actually turn over and dip to the east.

The rock is thus divided into three sections, sepa-

rated from each other by places of dislocation, where the
strata are much wrenched and broken.   These are
called the Northern and Southern Quebrada, from the
Spanish ' Tierra Quebrada,' or broken ground.   It is at
these places that the inland caves of Gibraltar are
almost exclusively found.   Based on the observations of
Dr. Falconer and himself, an excellent and most interesting account of these caves, and of the human
remains and works of art which they contain, was communicated by Mr. Busk to the meeting of the Congress
of Prehistoric Archæology at Norwich, and afterwards
printed in the ' Transactions ' of the Congress.[1]   Long
subsequent to the operation of the twisting force just
referred to, the promontory underwent various changes
of level.   There are sea-terraces and layers of shell-
breccia along its flanks, and numerous caves which, unlike
the inland ones, are the product of marine erosion.   The
Ape's Hill, on the African side of the strait, Mr. Busk
informs me has undergone similar disturbances.[2]

In the harbour of Gibraltar, on the morning of our
departure, I resumed a series of observations on the
colour of the sea.   On the way out a number of
specimens had been collected, with a view to subsequent
examination.   But the bottles were claret bottles, of
doubtful purity.   At Gibraltar, therefore, I purchased
fifteen white glass bottles, with ground glass stoppers,
and at Cadiz, thanks to the friendly guidance of Mr.
Cameron, I secured a dozen more.   These seven-and-

[1] In this essay Mr. Busk refers to the previous labours of Mr.
Smith, of Jordan Hill, to whom we owe most of our knowledge of
the geology of the rock.

[2] No one can rise from the perusal of Mr. Busk's paper without
a feeling of admiration for the principal discoverer and indefatigable explorer of the Gibraltar caves, the late Captain Frederick
Brome.

twenty bottles were filled with water, taken at different places between Oran and Spithead.

And here let me express my warmest acknowledgments to Captain Henderson, the commander of H.M.S. 'Urgent,' who aided me in my observations in every possible way. Indeed, my thanks are due to all the officers for their unfailing courtesy and help. The captain placed at my disposal his own coxswain, an intelligent fellow named Thorogood, who skilfully attached a cord to each bottle, weighted it with lead, cast it into the sea, and, after three successive rinsings, filled it under my own eyes. The contact of jugs, buckets, or other vessels was thus avoided; and even the necessity of pouring out the water, afterwards, through the dirty London air.

The mode of examination applied to these bottles has been already described.[1] The liquid is illuminated by a powerfully condensed beam, its condition being revealed through the light scattered by its suspended particles. 'Care is taken to defend the eye from the access of all other light, and, thus defended, it becomes an organ of inconceivable delicacy.' Were water of uniform density perfectly free from suspended matter, it would, in my opinion, scatter no light at all. The track of a luminous beam could not be seen in such water. But 'an amount of impurity so infinitesimal as to be scarcely expressible in numbers, and the individual particles of which are so small as wholly to elude the microscope, may, when examined by the method alluded to, produce not only sensible, but striking, effects upon the eye.'

The results of the examination of nineteen bottles filled at various places between Gibraltar and Spithead are here tabulated:

[1] 'Floating Matter of the Air,' Art. 'Dust and Disease.'

12

No.	Locality	Colour of Sea	Appearance in Luminous Beam
1	Gibraltar Harbour . . .	Green . . .	Thick with fine particles
2	Two miles from Gibraltar	Clearer green .	Thick with very fine particles
3	Off Cabreta Point . . .	Bright green .	Still thick, but less so
4	Off Cabreta Point . . .	Black-indigo .	Much less thick, very pure
5	Off Tarifa	Undecided . .	Thicker than No. 4
6	Beyond Tarifa	Cobalt-blue .	Much purer than No. 5
7	Twelve miles from Cadiz .	Yellow-green .	Very thick
8	Cadiz Harbour	Yellow-green .	Exceedingly thick
9	Fourteen miles from Cadiz	Yellow-green .	Thick, but less so
10	Fourteen miles from Cadiz	Bright green .	Much less thick
11	Between Capes St. Mary and Vincent	Deep indigo .	Very little matter, very pure
12	Off the Burlings. . . .	Strong green .	Thick, with fine matter
13	Beyond the Burlings . .	Indigo . . .	Very little matter, pure
14	Off Cape Finisterre. . .	Undecided . .	Less pure
15	Bay of Biscay	Black-indigo .	Very little matter, very pure
16	Bay of Biscay	Indigo . . .	Very fine matter. Iridescent
17	Off Ushant	Dark green . .	A good deal of matter
18	Off St. Catherine's . . .	Yellow-green .	Exceedingly thick
19	Spithead	Green . . .	Exceedingly thick

Here we have three specimens of water, described as green, a clearer green, and bright green, taken in Gibraltar Harbour, at a point two miles from the harbour, and off Cabreta Point. The home examination showed the first to be thick with suspended matter, the second less thick, and the third still less thick. Thus the green brightened as the suspended matter diminished in amount.

Previous to the fourth observation our excellent navigating lieutenant, Mr. Brown, steered along the coast, thus avoiding the adverse current which sets in, through the Strait, from the Atlantic to the Mediterranean. He was at length forced to cross the boundary of the Atlantic current, which was defined with extraordinary sharpness. On the one side of it the water was a vivid green, on the other a deep blue. Standing at the bow of the ship, a bottle could be filled with blue water, while at the same moment a bottle cast from the stern could be filled with green water. Two bottles were secured, one on each side of this remarkable boundary. In the distance the Atlantic had the hue

called ultra-marine; but looked fairly down upon, it
was of almost inky blackness—black qualified by a
trace of indigo.

What change does the home examination here
reveal? In passing to indigo, the water becomes suddenly augmented in purity, the suspended matter
becoming suddenly less. Off Tarifa, the deep indigo
disappears, and the sea is undecided in colour. Accompanying this change, we have a rise in the quantity of
suspended matter. Beyond Tarifa, we change to cobalt-
blue, the suspended matter falling at the same time in
quantity. This water is distinctly purer than the
green. We approach Cadiz, and at twelve miles from
the city get into yellow-green water; this the London
examination shows to be thick with suspended matter.
The same is true of Cadiz harbour, and also of a point
fourteen miles from Cadiz in the homeward direction.
Here there is a sudden change from yellow-green to a
bright emerald-green, and accompanying the change
a sudden fall in the quantity of suspended matter.
Between Cape St. Mary and Cape St. Vincent the
water changes to the deepest indigo, a further diminution
of the suspended matter being the concomitant phenomenon.

We now reach the remarkable group of rocks called
the Burlings, and find the water between the shore and
the rocks a strong green; the home examination shows
it to be thick with fine matter. Fifteen or twenty miles
beyond the Burlings we come again into indigo water,
from which the suspended matter has in great part disappeared. Off Cape Finisterre, about the place where
the 'Captain' went down, the water becomes green, and
the home examination pronounces it to be thicker. Then
we enter the Bay of Biscay, where the indigo resumes
its power, and where the home examination shows the

greatly augmented purity of the water. A second specimen of water, taken from the Bay of Biscay, held in suspension fine particles of a peculiar kind ; the size of them was such as to render the water richly iridescent. It showed itself green, blue, or salmon-coloured, according to the direction of the line of vision. Finally, we come to our last two bottles, the one taken opposite St. Catherine's lighthouse, in the Isle of Wight, the other at Spithead. The sea at both these places was green, and both specimens, as might be expected, were pronounced by the home examination to be thick with suspended matter.

Two distinct series of observations are here referred to—the one consisting of direct observations of the colour of the sea, conducted during the voyage from Gibraltar to Portsmouth : the other carried out in the laboratory of the Royal Institution. And here it is to be noted that in the home examination I never knew what water was placed in my hands. The labels, with the names of the localities written upon them, had been tied up, all information regarding the source of the water being thus held back. The bottles were simply numbered, and not till all of them had been examined, and described, were the labels opened, and the locality and sea-colour corresponding to the various specimens ascertained. The home observations, therefore, must have been perfectly unbiassed, and they clearly establish the association of the green colour with fine suspended matter, and of the ultramarine colour, and more especially of the black-indigo hue of the Atlantic, with the comparative absence of such matter.

So much for mere observation; but what is the cause of the dark hue of the deep ocean? [1] A prelimi-

[1] A note, written to me on October 22, by my friend Canon Kingsley, contains the following reference to this point: 'I have

nary remark or two will clear our way towards an explanation. Colour resides in white light, appearing when any constituent of the white light is withdrawn. The hue of a purple liquid, for example, is immediately accounted for by its action on a spectrum. It cuts out the yellow and green, and allows the red and blue to pass through. The blending of these two colours produces the purple. But while such a liquid attacks with special energy the yellow and green, it enfeebles the whole spectrum. By increasing the thickness of the stratum we may absorb the whole of the light. The colour of a blue liquid is similarly accounted for. It first extinguishes the red; then, as the thickness augments, it attacks the orange, yellow, and green in succession; the blue alone finally remaining. But even it might be extinguished by a sufficient depth of the liquid.

And now we are prepared for a brief, but tolerably complete, statement of that action of sea-water upon light, to which it owes its darkness. The spectrum embraces three classes of rays—the thermal, the visual, and the chemical. These divisions overlap each other; the thermal rays are in part visual, the visual rays in part chemical, and *vice versâ*. The vast body of thermal rays lie beyond the red, being invisible. These rays are attacked with exceeding energy by water. They are absorbed close to the surface of the sea, and are the great agents in evaporation. At the same time the whole spectrum suffers enfeeblement; water attacks all its rays, but with different degrees of energy. Of the

never seen the Lake of Geneva, but I thought of the brilliant dazzling dark blue of the mid-Atlantic under the sunlight, and its black-blue under cloud, both so solid that one might leap off the sponson on to it without fear; this was to me the most wonderful thing which I saw on my voyages to and from the West Indies.'

visual rays, the red are first extinguished. As the solar
beam plunges deeper into the sea, orange follows red,
yellow follows orange, green follows yellow, and the
various shades of blue, where the water is deep enough,
follow green. Absolute extinction of the solar beam
would be the consequence if the water were deep and
uniform. If it contained no suspended matter, such
water would be as black as ink. A reflected glimmer of
ordinary light would reach us from its surface, as it
would from the surface of actual ink; but no light,
hence no colour, would reach us from the body of the
water.

In very clear and deep sea-water this condition is
approximately fulfilled, and hence the extraordinary
darkness of such water. The indigo, already referred
to, is, I believe, to be ascribed in part to the suspended
matter, which is never absent, even in the purest natural
water; and in part to the slight reflection of the light
from the limiting surfaces of strata of different densities. A modicum of light is thus thrown back to the
eye, before the depth necessary to absolute extinction
has been attained. An effect precisely similar occurs
under the moraines of glaciers. The ice here is ex-
ceptionally compact, and, owing to the absence of the
internal scattering common in bubbled ice, the light
plunges into the mass, where it is extinguished, the
perfectly clear ice presenting an appearance of pitchy
blackness.[1]

The green colour of the sea has now to be accounted
for; and here, again, let us fall back upon the sure
basis of experiment. A strong white dinner-plate had
a lead weight securely fastened to it. Fifty or sixty
yards of strong hempen line were attached to the plate.

[1] I learn from a correspondent that certain Welsh tarns, which
are reputed bottomless, have this inky hue.

My assistant, Thorogood, occupied a boat, fastened as usual to the davits of the ' Urgent,' while I occupied a second boat nearer the stern of the ship. He cast the plate as a mariner heaves the lead, and by the time it reached me it had sunk a considerable depth in the water. In all cases the hue of this plate was green. Even when the sea was of the darkest indigo, the green was vivid and pronounced. I could notice the gradual deepening of the colour as the plate sank, but at its greatest depth, even in indigo water, the colour was still a blue-green.[1]

Other observations confirmed this one. The ' Urgent' is a screw steamer, and right over the blades of the screw was an orifice called the screw-well, through which one could look from the poop down upon the screw. The surface-glimmer, which so pesters the eye, was here in a great measure removed. Midway down, a plank crossed the screw-well from side to side; on this I placed myself and observed the action of the screw underneath. The eye was rendered sensitive by the moderation of the light; and, to remove still further all disturbing causes, Lieutenant Walton had a sail and tarpaulin thrown over the mouth of the well. Underneath this I perched myself on the plank and watched the screw. In an indigo sea the play of colour was indescribably beautiful, and the contrast between the water, which had the screw-blades, and that which had the bottom of the ocean, as a background, was extraordinary. The one was of the most brilliant green, the other of the deepest ultramarine. The surface of the water above the screw-blade was always ruffled. Liquid lenses were thus formed, by which the coloured light was withdrawn from some places and concentrated upon

[1] In no case, of course, is the green pure, but a mixture of green and blue.

others, the water flashing with metallic lustre. The
screw-blades in this case played the part of the dinner-
plate in the former case, and there were other instances
of a similar kind. The white bellies of porpoises
showed the green hue, varying in intensity as the
creatures swung to and fro between the surface and
the deeper water. Foam, at a certain depth below the
surface, was also green. In a rough sea the light which
penetrated the summit of a wave sometimes reached the
eye, a beautiful green cap being thus placed upon the
wave, even in indigo water.

But how is this colour to be connected with the suspended particles? Thus. Take the dinner-plate which
showed so brilliant a green when thrown into indigo
water. Suppose it to diminish in size, until it reaches
an almost microscopic magnitude. It would still behave
substantially as the larger plate, sending to the eye its
modicum of green light. If the plate, instead of being
a large coherent mass, were ground to a powder sufficiently fine, and in this condition diffused through the
clear sea-water, it would also send green light to the eye.
In fact, the suspended particles which the home examination reveals, act in all essential particulars like the
plate, or like the screw-blades, or like the foam, or like
the bellies of the porpoises. Thus I think the greenness of the sea is physically connected with the matter
which it holds in suspension.

We reached Portsmouth on January 5, 1871. Then
ended a voyage which, though its main object was not
realised, has left behind it pleasant memories, both of
the aspects of nature and the kindliness of men.

VII

NIAGARA.[1]

IT is one of the disadvantages of reading books about natural scenery that they fill the mind with pictures, often exaggerated, often distorted, often blurred, and, even when well drawn, injurious to the freshness of first impressions. Such has been the fate of most of us with regard to the Falls of Niagara. There was little accuracy in the estimates of the first observers of the cataract. Startled by an exhibition of power so novel and so grand, emotion leaped beyond the control of the judgment, and gave currency to notions which have often led to disappointment.

A record of a voyage in 1535 by a French mariner named Jacques Cartier, contains, it is said, the first printed allusion to Niagara. In 1603 the first map of the district was constructed by a Frenchman named Champlain. In 1648 the Jesuit Rageneau, in a letter to his superior at Paris, mentions Niagara as 'a cataract of frightful height.'[2] In the winter of 1678 and 1679 the cataract was visited by Father Hennepin, and described in a book dedicated 'to the King of Great Britain.' He gives a drawing of the waterfall, which

[1] A Discourse delivered at the Royal Institution of Great Britain, April 4, 1873.

[2] From an interesting little book presented to me at Brooklyn by its author, Mr. Holly, some of these data are derived : Hennepin, Kalm, Bakewell, Lyell, Hall, and others I have myself consulted.

shows that serious changes have taken place since his
time.   He describes it as 'a great and prodigious
cadence of water, to which the universe does not offer a
parallel.' The height of the fall, according to Hennepin,
was more than 600 feet.   'The waters,' he says, 'which
fall from this great precipice do foam and boil in the
most astonishing manner, making a noise more terrible
than that of thunder.   When the wind blows to the
south its frightful roaring may be heard for more than
fifteen leagues.'   The Baron la Hontan, who visited
Niagara in 1687, makes the height 800 feet.   In 1721
Charlevois, in a letter to Madame de Maintenon, after
referring to the exaggerations of his predecessors, thus
states the result of his own observations : 'For my part,
after examining it on all sides, I am inclined to think
that we cannot allow it less than 140 or 150 feet,'—a
remarkably close estimate.   At that time, viz. a hundred
and fifty years ago, it had the shape of a horseshoe, and
reasons will subsequently be given for holding that this
has been always the form of the cataract, from its origin
to its present site.

As regards the noise of the fall, Charlevois declares
the accounts of his predecessors, which, I may say, are
repeated to the present hour, to be altogether extrava-
gant.   He is perfectly right.   The thunders of Niagara
are formidable enough to those who really seek them
at the base of the Horseshoe Fall; but on the banks of
the river, and particularly above the fall, its silence,
rather than its noise, is surprising.   This arises, in
part, from the lack of resonance; the surrounding
country being flat, and therefore furnishing no echoing
surfaces to reinforce the shock of the water.   The
resonance from the surrounding rocks causes the Swiss
Reuss at the Devil's Bridge, when full, to thunder more
loudly than the Niagara.

On Friday, November 1, 1872, just before reaching the village of Niagara Falls, I caught, from the railway train, my first glimpse of the smoke of the cataract. Immediately after my arrival I went with a friend to the northern end of the American Fall. It may be that my mood at the time toned down the impression produced by the first aspect of this grand cascade; but I felt nothing like disappointment, knowing, from old experience, that time and close acquaintanceship, the gradual interweaving of mind and nature, must power-fully influence my final estimate of the scene. After dinner we crossed to Goat Island, and, turning to the right, reached the southern end of the American Fall. The river is here studded with small islands. Crossing a wooden bridge to Luna Island, and clasping a tree which grows near its edge, I looked long at the cataract, which here shoots down the precipice like an avalanche of foam. It grew in power and beauty. The channel spanned by the wooden bridge was deep, and the river there doubled over the edge of the precipice, like the swell of a muscle, unbroken. The ledge here over-hangs, the water being poured out far beyond the base of the precipice. A space, called the Cave of the Winds, is thus enclosed between the wall of rock and the falling water.

Goat Island ends in a sheer dry precipice, which connects the American and Horseshoe Falls. Midway between both is a wooden hut, the residence of the guide to the Cave of the Winds, and from the hut a winding staircase, called Biddle's Stair, descends to the base of the precipice. On the evening of my arrival I went down this stair, and wandered along the bottom of the cliff. One well-known factor in the formation and retreat of the cataract was immediately observed. A thick layer of limestone formed the upper portion of the cliff. This rested upon a bed of soft shale, which extended round the base of the cataract. The violent recoil of the water against this yielding substance crumbles it away, undermining the ledge above, which, unsupported, eventually breaks off, and produces the observed recession.

At the southern extremity of the Horseshoe is a promontory, formed by the doubling back of the gorge excavated by the cataract, and into which it plunges. On the promontory stands a stone building, called the Terrapin Tower, the door of which had been nailed up because of the decay of the staircase within it. Through the kindness of Mr. Townsend, the superintendent of Goat Island, the door was opened for me. From this tower, at all hours of the day, and at some hours of the night, I watched and listened to the Horseshoe Fall. The river here is evidently much deeper than the American branch; and instead of bursting into foam where it quits the ledge, it bends solidly over, and falls in a continuous layer of the most vivid green. The tint is not uniform; long stripes of deeper hue alternating with bands of brighter colour. Close to the ledge over which the water rolls, foam is generated, the light falling upon which, and flashing back from it, is sifted in its passage to and fro, and changed from white to emerald-green. Heaps of superficial foam are also formed at intervals along the ledge, and are immediately drawn into long white striæ.[1] Lower down, the surface, shaken by the reaction from below, incessantly rustles into whiteness. The descent finally resolves itself into a rhythm, the water reaching the bottom of the fall in periodic gushes. Nor is the

---

[1] The direction of the wind with reference to the course of a ship may be inferred with accuracy from the foam-streaks on the surface of the sea.

spray uniformly diffused through the air, but is wafted through it in successive veils of gauze-like texture. From all this it is evident that beauty is not absent from the Horseshoe Fall, but majesty is its chief attribute. The plunge of the water is not wild, but deliberate, vast, and fascinating. From the Terrapin Tower, the adjacent arm of the Horseshoe is seen projected against the opposite one, midway down ; to the imagination, therefore, is left the picturing of the gulf into which the cataract plunges.

The delight which natural scenery produces in some minds is difficult to explain, and the conduct which it prompts can hardly be fairly criticised by those who have never experienced it. It seems to me a deduction from the completeness of the celebrated Thomas Young, that he was unable to appreciate natural scenery. 'He had really,' says Dean Peacock, 'no taste for life in the country ; he was one of those who thought that no one who was able to, live in London would be content to live elsewhere.' Well, Dr. Young, like Dr. Johnson, had a right to his delights; but I can understand a hesitation to accept them, high as they were, to the exclusion of

> That o'erflowing joy which Nature yields
> To her true lovers.

To all who are of this mind, the strengthening of desire on my part to see and know Niagara Falls, as far as it is possible for them to be seen and known, will be intelligible.

On the first evening of my visit, I met, at the head of Biddle's Stair, the guide to the Cave of the Winds. He was in the prime of manhood—large, well built, firm and pleasant in mouth and eye. My interest in the scene stirred up his, and made him communicative.

Turning to a photograph, he described, by reference to it, a feat which he had accomplished some time previously, and which had brought him almost under the green water of the Horseshoe Fall. ' Can you lead me there to-morrow?' I asked. He eyed me enquiringly, weighing, perhaps, the chances of a man of light build, and with grey in his whiskers, in such an undertaking. ' I wish,' I added, ' to see as much of the fall as can be seen, and where you lead I will endeavour to follow.' His scrutiny relaxed into a smile, and he said, ' Very well; I shall be ready for you to-morrow.'

On the morrow, accordingly, I came. In the hut at the head of Biddle's Stair I stripped wholly, and re-dressed according to instructions,—drawing on two pairs of woollen pantaloons, three woollen jackets, two pairs of socks, and a pair of felt shoes. Even if wet, my guide assured me that the clothes would keep me from being chilled; and he was right. A suit and hood of yellow oilcloth covered all. Most laudable precautions were taken by the young assistant who helped to dress me to keep the water out; but his devices broke down immediately when severely tested.

We descended the stair; the handle of a pitchfork doing, in my case, the duty of an alpenstock. At the bottom, the guide enquired whether we should go first to the Cave of the Winds, or to the Horseshoe, remarking that the latter would try us most. I decided on getting the roughest done first, and he turned to the left over the stones. They were sharp and trying. The base of the first portion of the cataract is covered with huge boulders, obviously the ruins of the limestone ledge above. The water does not distribute itself uniformly among these, but seeks out channels through which it pours torrentially. We passed some of these with wetted feet, but without difficulty. At

length we came to the side of a more formidable
current. My guide walked along its edge until he
reached its least turbulent portion. Halting, he said,
'This is our greatest difficulty; if we can cross here,
we shall get far towards the Horseshoe.'

He waded in. It evidently required all his strength
to steady him. The water rose above his loins, and it
foamed still higher. He had to search for footing,
amid unseen boulders, against which the torrent rose
violently. He struggled and swayed, but he struggled
successfully, and finally reached the shallower water at
the other side. Stretching out his arm, he said to me,
'Now come on.' I looked down the torrent, as it
rushed to the river below, which was seething with the
tumult of the cataract. De Saussure recommended
the inspection of Alpine dangers, with the view of
making them familiar to the eye before they are en-
countered; and it is a wholesome custom in places of
difficulty to put the possibility of an accident clearly
before the mind, and to decide beforehand what ought
to be done should the accident occur. Thus wound up
in the present instance, I entered the water. Even
where it was not more than knee-deep, its power was
manifest. As it rose around me, I sought to split the
torrent by presenting a side to it; but the insecurity
of the footing enabled it to grasp my loins, twist me
fairly round, and bring its impetus to bear upon my
back. Further struggle was impossible; and feeling
my balance hopelessly gone, I turned, flung myself
towards the bank just quitted, and was instantly, as
expected, swept into shallower water.

The oilcloth covering was a great incumbrance; it
had been made for a much stouter man, and, standing
upright after my submersion, my legs occupied the
centre of two bags of water. My guide exhorted me to

try again. Prudence was at my elbow, whispering dissuasion; but, taking everything into account, it appeared more immoral to retreat than to proceed. Instructed by the first misadventure, I once more entered the stream. Had the alpenstock been of iron it might have helped me; but, as it was, the tendency of the water to sweep it out of my hands rendered it worse than useless. I, however, clung to it by habit. Again the torrent rose, and again I wavered; but, by keeping the left hip well against it, I remained upright, and at length grasped the hand of my leader at the other side. He laughed pleasantly. The first victory was gained, and he enjoyed it. ' No traveller,' he said, ' was ever here before.' Soon afterwards, by trusting to a piece of drift-wood which seemed firm, I was again taken off my feet, but was immediately caught by a protruding rock.

We clambered over the boulders towards the thickest spray, which soon became so weighty as to cause us to stagger under its shock. For the most part nothing could be seen; we were in the midst of bewildering tumult, lashed by the water, which sounded at times like the cracking of innumerable whips. Underneath this was the deep resonant roar of the cataract. I tried to shield my eyes with my hands, and look upwards; but the defence was useless. The guide continued to move on, but at a certain place he halted, desiring me to take shelter in his lee, and observe the cataract. The spray did not come so much from the upper ledge, as from the rebound of the shattered water when it struck the bottom. Hence the eyes could be protected from the blinding shock of the spray, while the line of vision to the upper ledges remained to some extent clear. On looking upwards over the guide's shoulder I could see the water bending

over the ledge, while the Terrapin Tower loomed fitfully
through the intermittent spray-gusts. We were right
under the tower. A little farther on the cataract, after
its first plunge, hit a protuberance some way down,
and flew from it in a prodigious burst of spray ; through
this we staggered. We rounded the promontory on
which the Terrapin Tower stands, and moved, amid
the wildest commotion, along the arm of the Horseshoe, until the boulders failed us, and the cataract fell
into the profound gorge of the Niagara River.

Here the guide sheltered me again, and desired me
to look up; I did so, and could see, as before, the
green gleam of the mighty curve sweeping over the
upper ledge, and the fitful plunge of the water, as
the spray between us and it alternately gathered and
disappeared. An eminent friend of mine often speaks
of the mistake of those physicians who regard man's
ailments as purely chemical, to be met by chemical
remedies only. He contends for the psychological
element of cure. By agreeable emotions, he says,
nervous currents are liberated which stimulate blood,
brain, and viscera. The influence rained from ladies'
eyes enables my friend to thrive on dishes which would
kill him if eaten alone. A sanative effect of the same
order I experienced amid the spray and thunder of
Niagara. Quickened by the emotions there aroused,
the blood sped exultingly through the arteries, abolish-
ing introspection, clearing the heart of all bitterness,
and enabling one to think with tolerance, if not with
tenderness, on the most relentless and unreasonable foe.
Apart from its scientific value, and purely as a moral
agent, the play was worth the candle. My companion
knew no more of me than that I enjoyed the wildness
of the scene ; but as I bent in the shelter of his large
frame he said, 'I should like to see you attempting to

13

describe all this.' He rightly thought it indescribable. The name of this gallant fellow was Thomas Conroy.

We returned, clambering at intervals up and down, so as to catch glimpses of the most impressive portions of the cataract. We passed under ledges formed by tabular masses of limestone, and through some curious openings formed by the falling together of the summits of the rocks. At length we found ourselves beside our enemy of the morning. Conroy halted for a minute or two, scanning the torrent thoughtfully. I said that, as a guide, he ought to have a rope in such a place; but he retorted that, as no traveller had ever thought of coming there, he did not see the necessity of keeping a rope. He waded in. The struggle to keep himself erect was evident enough; he swayed, but recovered himself again and again. At length he slipped, gave way, did as I had done, threw himself towards the bank, and was swept into the shallows. Standing in the stream near its edge, he stretched his arm towards me. I retained the pitchfork handle, for it had been useful among the boulders. By wading some way in, the staff could be made to reach him, and I proposed his seizing it. 'If you are sure,' he replied, 'that, in case of giving way, you can maintain your grasp, then I will certainly hold you.' Remarking that he might count on this, I waded in, and stretched the staff to my companion. It was firmly grasped by both of us. Thus helped, though its onset was strong, I moved safely across the torrent. All danger ended here. We afterwards roamed sociably among the torrents and boulders below the Cave of the Winds. The rocks were covered with organic slime, which could not have been walked over with bare feet, but the felt shoes effectually prevented slipping. We reached the cave and entered it, first by a wooden way carried over the boulders, and then along a narrow ledge, to the point eaten deepest into the shale. When the wind is from the south, the falling water, I am told, can be seen tranquilly from this spot; but when we were there, a blinding hurricane of spray was whirled against us. On the evening of the same day, I went behind the water on the Canada side, which, after the experiences of the morning, struck me as an imposture.

Still even this latter is exciting to some nerves. Its effect upon himself is thus vividly described by Mr. Bakewell, jun.: 'On' turning a sharp angle of the rock, a sudden gust of wind met us, coming from the hollow between the fall and the rock, which drove the spray directly in our faces, with such force that in an instant we were wet through. When in the midst of this shower-bath the shock took away my breath: I turned back and scrambled over the loose stones to escape the conflict. The guide soon followed, and told me that I had passed the worst part. With that assurance I made a second attempt; but so wild and disordered was my imagination that when I had reached half way I could bear it no longer.' [1]

To complete my knowledge I desired to see the fall from the river below it, and long negotiations were necessary to secure the means of doing so. The only boat fit for the undertaking had been laid up for the winter; but this difficulty, through the kind intervention of Mr. Townsend, was overcome. The main one was to secure oarsmen sufficiently strong and skilful to urge the boat where I wished it to be taken. The son of the owner of the boat, a finely-built young fellow, but only twenty, and therefore not sufficiently hardened, was willing to go; and up the river, it was stated, there lived another man who could do anything with the

[1] 'Mag. of Nat. Hist.,' 1830, pp. 121, 122.

boat which strength and daring could accomplish. He came. His figure and expression of face certainly indicated extraordinary firmness and power. On Tuesday, November 5, we started, each of us being clad in oilcloth. The elder oarsman at once assumed a tone of authority over his companion, and struck immediately in amid the breakers below the American Fall. He hugged the cross freshets instead of striking out into the smoother water. I asked him why he did so, and he replied that they were directed outwards, not downwards. The struggle, however, to prevent the bow of the boat from being turned by them, was often very severe.

The spray was in general blinding, but at times it disappeared and yielded noble views of the fall. The edge of the cataract is crimped by indentations which exalt its beauty. Here and there, a little below the highest ledge, a secondary one juts out; the water strikes it and bursts from it in huge protuberant masses of foam and spray. We passed Goat Island, came to the Horseshoe, and worked for a time along its base, the boulders over which Conroy and myself had scrambled a few days previously lying between us and the cataract. A rock was before us, concealed and revealed at intervals, as the waves passed over it. Our leader tried to get above this rock, first on the outside of it. The water, however, was here in violent motion. The men struggled fiercely, the older one ringing out an incessant peal of command and exhortation to the younger. As we were just clearing the rock, the bow came obliquely to the surge; the boat was turned suddenly round and shot with astonishing rapidity down the river. The men returned to the charge, now trying to get up between the half-concealed rock and the boulders to the left. But the torrent set in strongly through this channel. The tugging was quick and violent, but we made little way. At length, seizing a rope, the principal oarsman made a desperate attempt to get upon one of the boulders, hoping to be able to drag the boat through the channel; but it bumped so violently against the rock, that the man flung himself back and relinquished the attempt.

We returned along the base of the American Fall, running in and out among the currents which rushed from it laterally into the river. Seen from below the American Fall is certainly exquisitely beautiful, but it is a mere frill of adornment to its nobler neighbour the Horseshoe. At times we took to the river, from the centre of which the Horseshoe Fall appeared especially magnificent. A streak of cloud across the neck of Mont Blanc can double its apparent height, so here the green summit of the cataract shining above the smoke of spray appeared lifted to an extraordinary elevation. Had Hennepin and La Hontan seen the fall from this position, their estimates of the height would have been perfectly excusable.

From a point a little way below the American Fall, a ferry crosses the river, in summer, to the Canadian side. Below the ferry is a suspension bridge for carriages and foot-passengers, and a mile or two lower down is the railway suspension bridge. Between ferry and bridge the river Niagara flows unruffled; but at the suspension bridge the bed steepens and the river quickens its motion. Lower down the gorge narrows, and the rapidity and turbulence increase. At the place called the 'Whirlpool Rapids' I estimated the width of the river at 300 feet, an estimate confirmed by the dwellers on the spot. When it is remembered that the drainage of nearly half a continent is compressed into this space, the impetuosity of the river's rush may be imagined. Had it not been for Mr. Bierstädt, the distinguished photographer of Niagara, I should have quitted the place without seeing these rapids; for this, and for his agreeable company to the spot, I have to thank him. From the edge of the cliff above the rapids, we descended, a little, I confess, to a climber's disgust, in an ' elevator,' because the effects are best seen from the water level.

Two kinds of motion are here obviously active, a motion of translation and a motion of undulation—the race of the river through its gorge, and the great waves generated by its collision with, and rebound from, the obstacles in its way. In the middle of the river the rush and tossing are most violent; at all events, the impetuous force of the individual waves is here most strikingly displayed. Vast pyramidal heaps leap incessantly from the river, some of them with such energy as to jerk their summits into the air, where they hang momentarily suspended in crowds of liquid spherules. The sun shone for a few minutes. At times the wind, coming up the river, searched and sifted the spray, carrying away the lighter drops, and leaving the heavier ones behind. Wafted in the proper direction, rainbows appeared and disappeared fitfully in the lighter mist. In other directions the common gleam of the sunshine from the waves and their shattered crests was exquisitely beautiful. The complexity of the action was still further illustrated by the fact, that in some cases, as if by the exercise of a local explosive force, the drops were shot radially from a particular centre, forming around it a kind of halo.

The first impression, and, indeed, the current explanation of these rapids is, that the central bed of the river is cumbered with large boulders, and that the jostling, tossing, and wild leaping of the water there, are due to its impact against these obstacles. I doubt this explanation. At all events, there is another sufficient reason to be taken into account. Boulders derived from the adjacent cliffs visibly cumber the sides of the river. Against these the water rises and sinks rhythmically but violently, large waves being thus produced. On the generation of each wave, there is an immediate compounding of the wave-motion with the river-motion. The ridges, which in still water would proceed in circular curves round the centre of disturbance, cross the river obliquely, and the result is that at the centre waves commingle, which have really been generated at the sides. In the first instance, we had a composition of wave-motion with river-motion; here we have the coalescence of waves with waves. Where crest and furrow cross each other, the motion is annulled; where furrow and furrow cross, the river is ploughed to a greater depth; and where crest and crest aid each other, we have that astonishing leap of the water which breaks the cohesion of the crests, and tosses them shattered into the air. From the water level the cause of the action is not so easily seen; but from the summit of the cliff the lateral generation of the waves, and their propagation to the centre, are perfectly obvious. If this explanation be correct, the phenomena observed at the Whirlpool Rapids form one of the grandest illustrations of the principle of *interference*. The Nile ‘cataract,’ Mr. Huxley informs me, offers more moderate examples of the same action.

At some distance below the Whirlpool Rapids we have the celebrated whirlpool itself. Here the river makes a sudden bend to the north-east, forming nearly a right angle with its previous direction. The water strikes the concave bank with great force, and scoops it incessantly away. A vast basin has been thus formed, in which the sweep of the river prolongs itself in gyratory currents. Bodies and trees which have come over the falls, are stated to circulate here for days without finding the outlet. From various points of the cliffs above, this is curiously hidden. The rush of the river into the whirlpool is obvious enough; and though you imagine the outlet must be visible, if one existed, you cannot find it. Turning, however, round the bend of the precipice to the north-east, the outlet comes into view.

The Niagara season was over; the chatter of sight-seers had ceased, and the scene presented itself as one of holy seclusion and beauty. I went down to the river's edge, where the weird loneliness seemed to increase. The basin is enclosed by high and almost precipitous banks—covered, at the time, with russet woods. A kind of mystery attaches itself to gyrating water, due perhaps to the fact that we are to some extent ignorant of the direction of its force. It is said that at certain points of the whirlpool, pine-trees are sucked down, to be ejected mysteriously elsewhere. The water is of the brightest emerald-green. The gorge through which it escapes is narrow, and the motion of the river swift though silent. The surface is steeply inclined, but it is perfectly unbroken. There are no lateral waves, no ripples with their breaking bubbles to raise a murmur; while the depth is here too great to allow the inequality of the bed to ruffle the surface. Nothing can be more beautiful than this sloping liquid mirror formed by the Niagara, in sliding from the whirlpool.

The green colour is, I think, correctly accounted for in the last Fragment. While crossing the Atlantic in 1872–73 I had frequent opportunities of testing the explanation there given. Looked properly down upon, there are portions of the ocean to which we should hardly ascribe a trace of blue ; at the most, a mere hint of indigo reaches the eye. The water, indeed, is practically black, and this is an indication both of its depth and of its freedom from mechanically suspended matter. In small thicknesses water is sensibly transparent to all kinds of light ; but, as the thickness increases, the rays of low refrangibility are first absorbed, and after them the other rays. Where, therefore, the water is very deep and very pure, all the colours are absorbed, and such water ought to appear black, as no light is sent from its interior to the eye. The approximation of the Atlantic Ocean to this condition is an indication of its extreme purity.

Throw a white-pebble into such water ; as it sinks it becomes greener and greener, and, before it disappears, it reaches a vivid blue-green. Break such a pebble into fragments, each of these will behave like the unbroken mass ; grind the pebble to powder, every particle will yield its modicum of green ; and if the particles be so fine as to remain suspended in the water, the scattered light will be a uniform green. Hence the greenness of shoal water. You go to bed with the black Atlantic around you. You rise in the morning, find it a vivid green, and correctly infer that you are crossing the bank of Newfoundland. Such water is found charged with fine matter in a state of mechanical suspension. The light from the bottom may sometimes come into play, but it is not necessary. A storm can render the water muddy, by rendering the particles too numerous and gross. Such a case occurred towards the close of my visit to Niagara. There had been rain and storm in the upper lake-regions, and the quantity of suspended matter brought down quite extinguished the fascinating green of the Horseshoe.

Nothing can be more superb than the green of the Atlantic waves, when the circumstances are favourable to the exhibition of the colour. As long as a wave remains unbroken no colour appears; but when the foam just doubles over the crest, like an Alpine snow-cornice, under the cornice we often see a display of the most exquisite green. It is metallic in its brilliancy. But the foam is necessary to its production. The foam is first illuminated, and it scatters the light in all directions; the light which passes through the higher portion of the wave alone reaches the eye, and gives to that portion its matchless colour. The folding of the wave, producing as it does a series of longitudinal protuberances and furrows which act like cylindrical lenses, introduces variations in the intensity of the light, and materially enhances its beauty.

We have now to consider the genesis and proximate destiny of the Falls of Niagara. We may open our way to this subject by a few preliminary remarks upon erosion. Time and intensity are the main factors of geologic change, and they are in a certain sense convertible. A feeble force acting through long periods, and an intense force acting through short ones, may produce approximately the same results. To Dr. Hooker I have been indebted for some specimens of stones, the first examples of which were picked up by Mr. Hackworth on the shores of Lyell's Bay, near Wellington, in New Zealand. They were described by Mr. Travers in the 'Transactions of the New Zealand Institute.' Unacquainted with their origin, you would certainly ascribe their forms to human workmanship. They resemble knives and spear-heads, being apparently chiselled off into facets, with as much attention to symmetry as if a tool, guided by human intelligence, had passed over them. But no human instrument has been brought to bear upon these stones. They have been wrought into their present shape by the wind-blown sand of Lyell's Bay. Two winds are dominant here, and they in succession urged the sand against opposite sides of the stone; every little particle of sand chipped away its infinitesimal bit of stone, and in the end sculptured these singular forms.[1]

The Sphynx of Egypt is nearly covered up by the sand of the desert. The neck of the Sphynx is partly cut across, not, as I am assured by Mr. Huxley, by ordinary weathering, but by the eroding action of the fine sand blown against it. In these cases Nature furnishes us with hints which may be taken advantage of in art; and this action of sand has been recently turned to extraordinary account in the United States. When in Boston, I was taken by my courteous and helpful friend, Mr. Josiah Quincey, to see the action of the sand-blast. A kind of hopper containing fine silicious

[1] 'These stones, which have a strong resemblance to works of human art, occur in great abundance, and of various sizes, from half-an-inch to several inches in length. A large number were exhibited showing the various forms, which are those of wedges, knives, arrow-heads, &c., and all with sharp cutting edges.

'Mr. Travers explained that, notwithstanding their artificial appearance, these stones were formed by the cutting action of the wind-driven sand, as it passed to and fro over an exposed boulder-bank. He gave a minute account of the manner in which the varieties of form are produced, and referred to the effect which the erosive action thus indicated would have on railway and other works executed on sandy tracts.

'Dr. Hector stated that although, as a group, the specimens on the table could not well be mistaken for artificial productions, still the forms are so peculiar, and the edges, in a few of them, so perfect, that if they were discovered associated with human works, there is no doubt that they would have been referred to the so-called "stone period." '—*Extracted from the Minutes of the Wellington Philosophical Society*, February 9, 1869.

sand was connected with a reservoir of compressed air, the pressure being variable at pleasure. The hopper ended in a long slit, from which the sand was blown. A plate of glass was placed beneath this slit, and caused to pass slowly under it; it came out perfectly depolished, with a bright opalescent glimmer, such as could only be produced by the most careful grinding. Every little particle of sand urged against the glass, having all its energy concentrated on the point of impact, formed there a little pit, the depolished surface consisting of innumerable hollows of this description.

But this was not all. By protecting certain portions of the surface, and exposing others, figures and tracery of any required form could be etched upon the glass. The figures of open iron-work could be thus copied; while wire-gauze placed over the glass produced a reticulated pattern. But it required no such resisting substance as iron to shelter the glass. The patterns of the finest lace could be thus reproduced; the delicate filaments of the lace itself offering a sufficient protection. All these effects have been obtained with a simple model of the sand-blast devised by my assistant. A fraction of a minute suffices to etch upon glass a rich and beautiful lace pattern. Any yielding substance may be employed to protect the glass. By diffusing the shock of the particle, such substances practically destroy the local erosive power. The hand can bear, without inconvenience, a sand-shower which would pulverise glass. Etchings executed on glass with suitable kinds of ink are accurately worked out by the sand-blast. In fact, within certain limits, the harder the surface, the greater is the concentration of the shock, and the more effectual is the erosion. It is not necessary that the sand should be the harder substance of the two; corundum, for example, is much harder than quartz; still, quartz-sand can not only depolish, but actually blow a hole through a plate of corundum. Nay, glass may be depolished by the impact of fine shot; the grains in this case bruising the glass, before they have time to flatten and turn their energy into heat.

And here, in passing, we may tie together one or two apparently unrelated facts. Supposing you turn on, at the lower part of a house, a cock which is fed by a pipe from a cistern at the top of the house, the column of water, from the cistern downwards, is set in motion. By turning off the cock, this motion is stopped; and when the turning off is very sudden, the pipe, if not strong, may be burst by the internal impact of the water. By distributing the turning of the cock over half a second of time, the shock and danger of rupture may be entirely avoided. We have here an example of the concentration of energy in *time*. The sand-blast illustrates the concentration of energy in *space*. The action of flint and steel is an illustration of the same principle. The heat required to generate the spark is intense; and the mechanical action, being moderate, must, to produce fire, be in the highest degree concentrated. This concentration is secured by the collision of hard substances. Calc-spar will not supply the place of flint, nor lead the place of steel, in the production of fire by collision. With the softer substances, the total heat produced may be greater than with the hard ones, but, to produce the spark, the heat must be intensely localised.

We can, however, go far beyond the mere depolishing of glass; indeed I have already said that quartz-sand can wear a hole through corundum. This leads me to express my acknowledgments to General Tilghman,[1] who

[1] The absorbent power, if I may use the phrase, exerted by the industrial arts in the United States, is forcibly illustrated by the

is the inventor of the sand-blast. To his spontaneous
kindness I am indebted for some beautiful illustrations
of his process. In one thick plate of glass a figure has
been worked out to a depth of $\frac{3}{8}$ths of an inch. A
second plate, $\frac{7}{8}$ths of an inch thick, is entirely per-
forated. In a circular plate of marble, nearly half
an inch thick, open work of most intricate and
elaborate description has been executed. It would probably take many days to perform this work by any
ordinary process; with the sand-blast it was accomplished in an hour. So much for the strength of the
blast; its delicacy is illustrated by this beautiful
example of line engraving, etched on glass by means
of the blast.

This power of erosion, so strikingly displayed when
sand is urged by air, renders us better able to conceive
its action when urged by water. The erosive power of
a river is vastly augmented by the solid matter carried
along with it. Sand or pebbles, caught in a river
vortex, can wear away the hardest rock; 'potholes' and
deep cylindrical shafts being thus produced. An extraordinary instance of this kind of erosion is to be seen
in the Val Tournanche, above the village of this name.
The gorge at Handeck has been thus cut out. Such
waterfalls were once frequent in the valleys of Switzerland; for hardly any valley is without one or more
transverse barriers of resisting material, over which the
river flowing through the valley once fell as a cataract.
Near Pontresina, in the Engadin, there is such a case;

rapid transfer of men like Mr. Tilghman from the life of the soldier
to that of the civilian. General McClellan, now a civil engineer,
whom I had the honour of frequently meeting in New York, is a
most eminent example of the same kind. At the end of the war,
indeed, a million and a half of men were thus drawn, in an as-
tonishingly short time, from military to civil life.

a hard gneiss being there worn away to form a gorge, through which the river from the Morteratsch glacier rushes. The barrier of the Kirchet above Meyringen is also a case in point. Behind it was a lake, derived from the glacier of the Aar, and over the barrier the lake poured its excess of water. Here the rock, being limestone, was in part dissolved; but added to this we had the action of the sand and gravel carried along by the water, which, on striking the rock, chipped it away like the particles of the sand-blast. Thus, by solution and mechanical erosion, the great chasm of the Finsteraarschlucht was formed. It is demonstrable that the water which flows at the bottoms of such deep fissures once flowed at the level of their present edges, and tumbled down the lower faces of the barriers. Almost every valley in Switzerland furnishes examples of this kind; the untenable hypothesis of earthquakes, once so readily resorted to in accounting for these gorges, being now for the most part abandoned. To produce the Cañons of Western America, no other cause is needed than the integration of effects individually infinitesimal.

And now we come to Niagara. Soon after Europeans had taken possession of the country, the conviction appears to have arisen that the deep channel of the river Niagara below the falls had been excavated by the cataract. In Mr. Bakewell's 'Introduction to Geology,' the prevalence of this belief has been referred to. It is expressed thus by Professor Joseph Henry in the 'Transactions of the Albany Institute:'[1] ' In viewing the position of the falls, and the features of the country round, it is impossible not to be impressed with the idea that this great natural raceway has been formed

[1] Quoted by Bakewell.

by the continued action of the irresistible Niagara, and that the falls, beginning at Lewiston, have, in the course of ages, worn back the rocky strata to their present site.' The same view is advocated by Sir Charles Lyell, by Mr. Hall, by M. Agassiz, by Professor Ramsay, indeed by most of those who have inspected the place.

A connected image of the origin and progress of the cataract is easily obtained. Walking northward from the village of Niagara Falls by the side of the river, we have to our left the deep and comparatively narrow gorge, through which the Niagara flows. The bounding cliffs of this gorge are from 300 to 350 feet high. We reach the whirlpool, trend to the north-east, and after a little time gradually resume our northward course. Finally, at about seven miles from the present falls, we come to the edge of a declivity, which informs us that we have been hitherto walking on table-land. At some hundreds of feet below us is a comparatively level plain, which stretches to Lake Ontario. The declivity marks the end of the precipitous gorge of the Niagara. Here the river escapes from its steep mural boundaries, and in a widened bed pursues its way to the lake which finally receives its waters.

The fact that in historic times, even within the memory of man, the fall has sensibly receded, prompts the question, How far has this recession gone? At what point did the ledge which thus continually creeps backwards begin its retrograde course? To minds disciplined in such researches the answer has been, and will be—At the precipitous declivity which crossed the Niagara from Lewiston on the American to Queenston on the Canadian side. Over this transverse barrier the united affluents of all the upper lakes once poured their waters, and here the work of erosion began. The dam, moreover, was demonstrably of sufficient height to cause the river above it to submerge Goat Island ; and this would perfectly account for the finding by Sir Charles Lyell, Mr. Hall, and others, in the sand and gravel of the island, the same fluviatile shells as are now found in the Niagara River higher up. It would also account for those deposits along the sides of the river, the discovery of which enabled Lyell, Hall, and Ramsay to reduce to demonstration the popular belief that the Niagara once flowed through a shallow valley.

The physics of the problem of excavation, which I made clear to my mind before quitting Niagara, are revealed by a close inspection of the present Horseshoe Fall. We see evidently that the greatest weight of water bends over the very apex of the Horseshoe. In a passage in his excellent chapter on Niagara Falls, Mr. Hall alludes to this fact. Here we have the most copious and the most violent whirling of the shattered liquid ; here the most powerful eddies recoil against the shale. From this portion of the fall, indeed, the spray sometimes rises without solution of continuity to the region of clouds, becoming gradually more attenuated, and passing finally through the condition of true cloud into invisible vapour, which is sometimes reprecipitated higher up. All the phenomena point distinctly to the centre of the river as the place of greatest mechanical energy, and from the centre the vigour of the fall gradually dies away towards the sides. The Horseshoe form, with the concavity facing downwards, is an obvious and necessary consequence of this action. Right along the middle of the river the apex of the curve pushes its way backwards, cutting along the centre a deep and comparatively narrow groove, and draining the

14

sides as it passes them.[1]   Hence the remarkable discrepancy between the widths of the Niagara above and
below the Horseshoe.   All along its course, from Lewiston Heights to its present position, the form of the fall
was probably that of a horseshoe; for this is merely
the expression of the greater depth, and consequently
greater excavating power, of the centre of the river.
The gorge, moreover, varies in width, as the depth of
the centre of the ancient river varied, being narrowest
where that depth was greatest.

The vast comparative erosive energy of the Horseshoe Fall comes strikingly into view when it and the
American Fall are compared together.   The American
branch of the river is cut at a right angle by the
gorge of the Niagara.   Here the Horseshoe Fall was
the real excavator.   It cut the rock, and formed the
precipice, over which the American Fall tumbles.   But
since its formation, the erosive action of the American
Fall has been almost nil, while the Horseshoe has cut
its way for 500 yards across the end of Goat Island, and
is now doubling back to excavate its channel parallel to
the length of the island.   This point, which impressed
me forcibly, has not, I have just learned, escaped the
acute observation of Professor Ramsay.[2]   The river
bends; the Horseshoe immediately accommodates itself to the bending, and will follow implicitly the direction of the deepest water in the upper stream.   The

[1] In the discourse the excavation of the centre and drainage of
the sides action was illustrated by a model devised by my assistant,
Mr. John Cottrell.

[2] His words are: 'Where the body of water is small in the
American Fall, the edge has only receded a few yards (where most
eroded) during the time that the Canadian Fall has receded from
the north corner of Goat Island to the innermost curve of the
Horseshoe Fall.'—*Quarterly Journal of Geological Society*, May
1859.

flexures of the gorge are determined by those of the river channel above it. Were the Niagara centre above the fall sinuous, the gorge would obediently follow its sinuosities. Once suggested, no doubt geographers will be able to point out many examples of this action. The Zambesi is thought to present a great difficulty to the erosion theory, because of the sinuosity of the chasm below the Victoria Falls. But, assuming the basalt to be of tolerably uniform texture, had the river been examined before the formation of this sinuous channel, the present zigzag course of the gorge below the fall could, I am persuaded, have been predicted, while the sounding of the present river would enable us to predict the course to be pursued by the erosion in the future.

But not only has the Niagara River cut the gorge; it has carried away the chips of its own workshop. The shale, being probably crumbled, is easily carried away. But at the base of the fall we find the huge boulders already described, and by some means or other these are removed down the river. The ice which fills the gorge in winter, and which grapples with the boulders, has been regarded as the transporting agent. Probably it is so to some extent. But erosion acts without ceasing on the abutting points of the boulders, thus withdrawing their support and urging them gradually down the river. Solution also does its portion of the work. That solid matter is carried down is proved by the difference of depth between the Niagara River and Lake Ontario, where the river enters it. The depth falls from 72 feet to 20 feet, in consequence of the deposition of solid matter caused by the diminished motion of the river.[1]

[1] Near the mouth of the gorge at Queenston, the depth, according to the Admiralty Chart, is 180 feet; well within the gorge it is 132 feet.

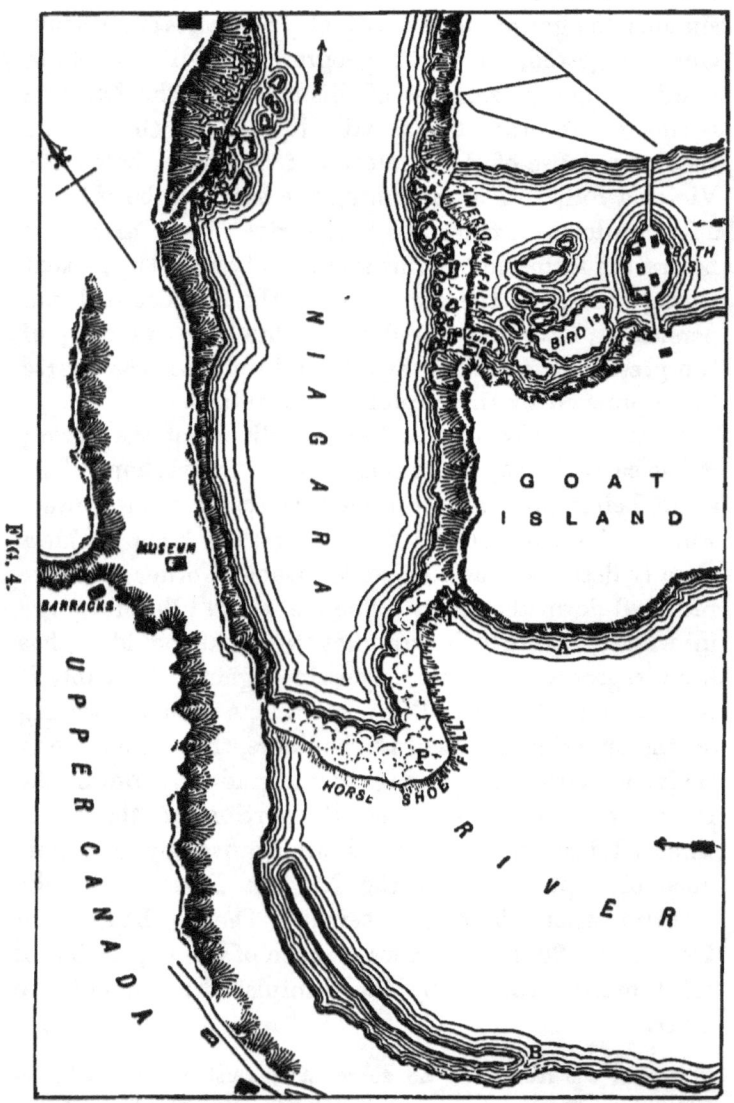

FIG. 4.

The annexed highly instructive map has been reduced from one published in Mr. Hall's 'Geology of
New York.' It is based on surveys executed in 1842, by
Messrs. Gibson and Evershed. The ragged edge of the
American Fall north of Goat Island marks the amount
of erosion which it has been able to accomplish, while
the Horseshoe Fall was cutting its way southward across
the end of Goat Island to its present position. The
American Fall is 168 feet high, a precipice cut down,
not by itself, but by the Horseshoe Fall. The latter in
1842 was 159 feet high, and, as shown by the map, is
already turning eastward, to excavate its gorge along
the centre of the upper river. P is the apex of the
Horseshoe, and T marks the site of the Terrapin Tower,
with the promontory adjacent, round which I was conducted by Conroy Probably since 1842 the Horseshoe has worked back beyond the position here assigned
to it.

In conclusion, we may say a word regarding the
proximate future of Niagara. At the rate of excavation
assigned to it by Sir Charles Lyell, namely, a foot a year,
five thousand years or so will carry the Horseshoe Fall
far higher than Goat Island. As the gorge recedes it
will drain, as it has hitherto done, the banks right and
left of it, thus leaving a nearly level terrace between
Goat Island and the edge of the gorge. Higher up it
will totally drain the American branch of the river ; the
channel of which in due time will become cultivable
land. The American Fall will then be transformed into
a dry precipice, forming a simple continuation of the
cliffy boundary of the Niagara gorge. At the place
occupied by the fall at this moment we shall have the
gorge enclosing a right angle, a second whirlpool being
the consequence. To those who visit Niagara a few
millenniums hence I leave the verification of this pre-

diction. All that can be said is, that if the causes now in action continue to act, it will prove itself literally true.

POSTSCRIPT.

A year or so after I had quitted the United States, a man sixty years of age, while engaged in painting one of the bridges which connect Goat Island with the Three Sisters, slipped through the rails of the bridge into the rapids, and was carried impetuously towards the Horseshoe Fall. He was urged against a rock which rose above the water, and with the grasp of desperation he clung to it. The population of the village of Niagara Falls was soon upon the island, and ropes were brought, but there was none to use them. In the midst of the excitement, a tall powerful young fellow was observed making his way silently through the crowd. He reached a rope; selected from the bystanders a number of men, and placed one end of the rope in their hands. The other end he fastened round himself, and choosing a point considerably above that to which the man clung, he plunged into the rapids. He was carried violently downwards, but he caught the rock, secured the old painter and saved him. Newspapers from all parts of the Union poured in upon me, describing this gallant act of my guide Conroy.

VIII

THE PARALLEL ROADS OF GLEN ROY.[1]

THE first published allusion to the Parallel Roads of Glen Roy occurs in the appendix to the third volume of Pennant's ' Tour in Scotland,' a work published in 1776. ' In the face of these hills,' says this writer, ' both sides of the glen, there are three roads at small distances from each other and directly opposite on each side. These roads have been measured in the complete parts of them; and found to be 26 paces of a man 5 feet 10 inches high. The two highest are pretty near each other, about 50 yards, and the lowest double that distance from the nearest to it. They are carried along the sides of the glen with the utmost regularity, nearly as exact as drawn with a line of rule and compass.'

The correct heights of the three roads of Glen Roy are respectively 1150, 1070, and 860 feet above the sea. Hence a vertical distance of 80 feet separates the two highest, while the lowest road is 210 feet below the middle one.

These ' roads ' are usually shelves or terraces formed in the yielding drift which here covers the slopes of the mountains. They are all sensibly horizontal and therefore parallel. Pennant accepted as reasonable the explanation of them given by the country people in his

[1] A discourse delivered at the Royal Institution of Great Britain on June 9, 1876.

time.   They thought that the roads 'were designed for
the chase, and that the terraces were made after the
spots were cleared in lines from wood, in order to tempt
the animals into the open paths after they were rouzed,
in order that they might come within reach of the
bowmen who might conceal themselves in the woods
above and below.'

In these attempts of 'the country people' we have
an illustration of that impulse to which all scientific
knowledge is due—the desire to know the causes of
things ; and it is a matter of surprise that in the case of
the parallel roads, with their weird appearance chal-
lenging enquiry, this impulse did not make itself more
rapidly and energetically felt.   Their remoteness may
perhaps account for the fact that until the year 1817
no systematic description of them, and no scientific
attempt at an explanation of them, appeared.   In that
year Dr. MacCulloch, who was then President of the
Geological Society, presented to that Society a memoir,
in which the roads were discussed, and pronounced to
be the margins of lakes once embosomed in Glen Roy.
Why there should be three roads, or why the lakes
should stand at these particular levels, was left unexplained.

To Dr. MacCulloch succeeded a man, possibly not so
learned as a geologist, but obviously fitted by nature to
grapple with her facts and to put them in their
proper setting.   I refer to Sir Thomas Dick-Lauder,
who presented to the Royal Society of Edinburgh, on
the 2nd of March, 1818, his paper on the Parallel Roads
of Glen Roy.   In looking over the literature of this
subject, which is now copious, it is interesting to observe
the differentiation of minds, and to single out those
who went by a kind of instinct to the core of the question, from those who erred in it, or who learnedly

occupied themselves with its analogies, adjuncts, and details. There is no man, in my opinion, connected with the history of the subject, who has shown, in relation to it, this spirit of penetration, this force of scientific insight, more conspicuously than Sir Thomas Dick-Lauder. Two distinct mental processes are involved in the treatment of such a question. Firstly, the faithful and sufficient observation of the data; and secondly, that higher mental process in which the constructive imagination comes into play, connecting the separate facts of observation with their common cause, and weaving them into an organic whole. In neither of these requirements did Sir Thomas Dick-Lauder fail.

Adjacent to Glen Roy is a valley called Glen Gluoy, along the sides of which ran a single shelf, or terrace, formed obviously in the same manner as the parallel roads of Glen Roy. The two shelves on the opposing sides of the glen were at precisely the same level, and Dick-Lauder wished to see whether, and how, they became united at the head of the glen. He followed the shelves into the recesses of the mountains. The bottom of the valley, as it rose, came ever nearer to them, until finally, at the head of Glen Gluoy, he reached a col, or watershed, of precisely the same elevation as the road which swept round the glen.

The correct height of this col is 1170 feet above the sea; that is to say, 20 feet above the highest road in Glen Roy.

From this col a lateral branch-valley—Glen Turrit —led down to Glen Roy. Our explorer descended from the col to the highest road of the latter glen, and pursued it exactly as he had pursued the road in Glen Gluoy. For a time it belted the mountain sides at a considerable height above the bottom of the valley; but this rose as he proceeded, coming ever nearer to the

highest shelf, until finally he reached a col, or watershed, looking into Glen Spey, and of precisely the same
elevation as the highest road of Glen Roy.

PARALLEL ROADS OF GLEN ROY.

After a Sketch by SIR THOMAS DICK-LAUDER.

He then dropped down to the lowest of these roads,
and followed it towards the mouth of the glen. Its
elevation above the bottom of the valley gradually
increased ; not because the shelf rose, but because it
remained level while the valley sloped downwards. He
found this lowest road doubling round the hills at the
mouth of Glen Roy, and running along the sides of the
mountains which flank Glen Spean. He followed it
eastwards. The bottom of the Spean Valley, like the
others, gradually rose, and therefore gradually approached the road on the adjacent mountain-side. He
came to Loch Laggan, the surface of which rose almost

to the level of the road, and beyond the head of this lake ne found, as in the other two cases, a col, or watershed, at Makul, of exactly the same level as the single road in Glen Spean, which, it will be remembered, is a continuation of the lowest road in Glen Roy.

Here we have a series of facts of obvious significance as regards the solution of this problem. The effort of the mind to form a coherent image from such facts may be compared with the effort of the eyes to cause the pictures of a stereoscope to coalesce. For a time we exercise a certain strain, the object remaining vague and indistinct. Suddenly its various parts seem to run together, the object starting forth in clear and definite relief. Such, I take it, was the effect of his ponderings upon the mind of Sir Thomas Dick-Lauder. His solution was this : Taking all their features into account, he was convinced that water only could have produced the terraces. But how had the water been collected? He saw clearly that, supposing the mouth of Glen Gluoy to be stopped by a barrier sufficiently high, if the waters from the mountains flanking the glen were allowed to collect, they would form behind the barrier a lake, the surface of which would gradually rise until it reached the level of the col at the head of the glen. The rising would then cease ; the superfluous water of Glen Gluoy discharging itself over the col into Glen Roy. As long as the barrier stopping the mouth of Glen Gluoy continued high enough, we should have in that glen a lake at the precise level of its shelf, which lake, acting upon the loose drift of the flanking mountains, would form the shelf revealed by observation.

So much for Glen Gluoy. But suppose the mouth of Glen Roy also stopped by a similar barrier. Behind it also the water from the adjacent mountains would collect. The surface of the lake thus formed would

gradually rise, until it had reached the level of the
col which divides Glen Roy from Glen Spey. Here
the rising of the lake would cease; its superabundant
water being poured over the col into the valley of the
Spey. This state of things would continue as long as
a sufficiently high barrier remained at the mouth of
Glen Roy. The lake thus dammed in, with its surface
at the level of the highest parallel road, would act, as in
Glen Gluoy, upon the friable drift overspreading the
mountains, and would form the highest road or terrace
of Glen Roy.

And now let us suppose the barrier to be so far
removed from the mouth of Glen Roy as to establish a
connection between it and the upper part of Glen Spean,
while the lower part of the latter glen still continued
to be blocked up. Upper Glen Spean and Glen Roy
would then be occupied by a continuous lake, the level
of which would obviously be determined by the col at
the head of Loch Laggan. The water in Glen Roy
would sink from the level it had previously maintained,
to the level of its new place of escape. This new lake-
surface would correspond exactly with the lowest parallel
road, and it would form that road by its action upon the
drift of the adjacent mountains.

In presence of the observed facts, this solution com-
mends itself strongly to the scientific mind. The
question next occurs, What was the character of the
assumed barrier which stopped the glens? There are
at the present moment vast masses of detritus in certain
portions of Glen Spean, and of such detritus Sir
Thomas Dick-Lauder imagined his barriers to have
been formed. By some unknown convulsion, this
detritus had been heaped up. But, once given, and
once granted that it was subsequently removed in the
manner indicated, the single road of Glen Gluoy and

the highest and lowest roads of Glen Roy would be explained in a satisfactory manner.

To account for the second or middle road of Glen
Roy, Sir Thomas Dick-Lauder invoked a new agency.
He supposed that at a certain point in the breaking
down or waste of his dam, a halt occurred, the barrier
holding its ground at a particular level sufficiently long
to dam a lake rising to the height of, and forming the
second road. This point of weakness was at once detected by Mr. Darwin, and adduced by him as proving
that the levels of the cols did not constitute an essential
feature in the phenomena of the parallel roads. Though
not destroyed, Sir Thomas Dick-Lauder's theory was
seriously shaken by this argument, and it became a
point of capital importance, if the facts permitted, to
remove such source of weakness. This was done in
1847 by Mr. David Milne, now Mr. Milne-Home. On
walking up Glen Roy from Roy Bridge, we pass the
mouth of a lateral glen, called Glen Glaster, running
eastward from Glen Roy. There is nothing in this
lateral glen to attract attention, or to suggest that it
could have any conspicuous influence in the production
of the parallel roads. Hence, probably, the failure of
Sir Thomas Dick-Lauder to notice it. But Mr. Milne-
Home entered this glen, on the northern side of which
the middle and lowest roads are fairly shown. The
principal stream running through the glen turns at a
certain point northwards and loses itself among hills
too high to offer any outlet. But another branch of
the glen turns to the south-east ; and, following up
this branch, Mr. Milne-Home reached a col, or watershed, of the precise level of the second Glen Roy road.
When the barrier blocking the glens had been so far
removed as to open this col, the water in Glen Roy
would sink to the level of the second road. A new

lake of diminished depth would be thus formed, the
surplus water of which would escape over the Glen
Glaster col into Glen Spean. The margin of this new
lake, acting upon the detrital matter, would form the
second road. The theory of Sir Thomas Dick-Lauder,
as regards the part played by the cols, was re-riveted
by this new and unexpected discovery.

I have referred to Mr. Darwin, whose powerful
mind swayed for a time the convictions of the scientific
world in relation to this question. His notion was—
and it is a notion which very naturally presents itself—
that the parallel roads were formed by the sea; that
this whole region was once submerged and subsequently
upheaved; that there were pauses in the process of upheaval, during which these glens constituted so many
fiords, on the sides of which the parallel terraces were
formed. This theory will not bear close criticism; nor
is it now maintained by Mr. Darwin himself. It would
not account for the sea being 20 feet higher in Glen
Gluoy than in Glen Roy. It would not account for the
absence of the second and third Glen Roy roads from
Glen Gluoy, where the mountain flanks are quite as im-
pressionable as in Glen Roy. It would not account for
the absence of the shelves from the other mountains in
the neighbourhood, all of which would have been
clasped by the sea had the sea been there. Here then,
and no doubt elsewhere, Mr. Darwin has shown himself
to be fallible; but here, as elsewhere, he has shown
himself equal to that discipline of surrender to evidence
which girds his intellect with such unassailable moral
strength.   -

But, granting the significance of Sir Thomas Dick-
Lauder's facts, and the reasonableness, on the whole, of
the views which he has founded on them, they will not
bear examination in detail. No such barriers of

detritus as he assumed could have existed without leaving traces behind them; but there is no trace left. There is detritus enough in Glen Spean, but not where it is wanted.   The two highest parallel roads stop abruptly at different points near the mouth of Glen Roy, but no remnant of the barrier against which they abutted is to be seen.   It might be urged that the subsequent invasion of the valley by glaciers has swept the detritus away; but there have been no glaciers in these valleys since the disappearance of the lakes. Professor Geikie has favoured me with a drawing of the Glen Spean 'road' near the entrance to Glen Trieg.   The road forms a shelf round a great mound of detritus which, had a glacier followed the formation of the shelf, must have been cleared away.   Taking all the circumstances into account, you may, I think, with safety dismiss the detrital barrier as incompetent to account for the present condition of Glen Gluoy and Glen Roy.

Hypotheses in science, though apparently transcending experience, are in reality experience modified by scientific thought and pushed into an ultra experiential region.   At the time that he wrote, Sir Thomas Dick-Lauder could not possibly have discerned the cause subsequently assigned for the blockage of these glens.   A knowledge of the action of ancient glaciers was the necessary antecedent to the new explanation, and experience of this nature was not possessed by the distinguished writer just mentioned.   The extension of Swiss glaciers far beyond their present limits, was first made known by a Swiss engineer named Venetz, who established, by the marks they had left behind them, their former existence in places which they had long forsaken.   The subject of glacier extension was subsequently followed up with distinguished success by

Charpentier, Studer, and others.   With characteristic
vigour Agassiz grappled with it, extending his observations far beyond the domain of Switzerland.   He
came to this country in 1840, and found in various
places indubitable marks of ancient glacier action.
England, Scotland, Wales, and Ireland he proved to
have once given birth to glaciers.   He visited Glen
Roy, surveyed the surrounding neighbourhood, and
pronounced, as a consequence of his investigation, the
barriers which stopped the glens and produced the
parallel roads to have been barriers of ice.   To Mr.
Jamieson, above all others, we are indebted for the
thorough testing and confirmation of this theory.

And let me here say that Agassiz is only too likely
to be misrated and misjudged by those who, though
accurate within a limited sphere, fail to grasp in their
totality the motive powers invoked in scientific investigation.   True he lacked mechanical precision, but he
abounded in that force and freshness of the scientific
imagination which in some sciences, and probably in
some stages of all sciences, are essential to the creator
of knowledge.   To Agassiz was given, not the art of
the refiner, but the instinct of the discoverer, and the
strength of the delver who brings ore from the recesses
of the mine.   That ore may contain its share of dross,
but it also contains the precious metal which gives
employment to the refiner, and without which his
occupation would depart.

Let us dwell for a moment upon this subject of
ancient glaciers.   Under a flask containing water, in
which a thermometer is immersed, is placed a Bunsen's
lamp.   The water is heated, reaches a temperature of
212°, and then begins to boil.   The rise of the thermometer then ceases, although heat continues to be
poured by the lamp into the water.   What becomes of

that heat? We know that it is consumed in the molecular work of vaporization. In the experiment here arranged, the steam passes from the flask through a tube into a second vessel kept at a low temperature. Here it is condensed, and indeed congealed to ice, the second vessel being plunged in a mixture cold enough to freeze the water. As a result of the process we obtain a mass of ice. That ice has an origin very antithetical to its own character. Though cold, it is the child of heat. If we removed the lamp, there would be no steam, and if there were no steam there would be no ice. The mere cold of the mixture surrounding the second vessel would not produce ice. The cold must have the proper material to work upon; and this material—aqueous vapour—is, as we here see, the direct product of heat.

It is now, I suppose, fifteen or sixteen years since I found myself conversing with an illustrious philosopher regarding that glacial epoch which the researches of Agassiz and others had revealed. This profoundly thoughtful man maintained the fixed opinion that, at a certain stage in the history of the solar system, the sun's radiation had suffered diminution, the glacial epoch being a consequence of this solar chill. The celebrated French mathematician Poisson had another theory. Astronomers have shown that the solar system moves through space, and 'the temperature of space' is a familiar expression with scientific men. It was considered probable by Poisson that our system, during its motion, had traversed portions of space of different temperatures; and that, during its passage through one of the colder regions of the universe, the glacial epoch occurred. Notions such as these were more or less current everywhere not many years ago, and I therefore thought it worth while to show how incom-

15

plete they were.   Suppose the temperature of our
planet to be reduced, by the subsidence of solar heat,
the cold of space, or any other cause, say one hundred
degrees.   Four-and-twenty hours of such a chill would
bring down as snow nearly all the moisture of our
atmosphere.   But this would not produce a glacial
epoch.   Such an epoch would require the long-continued
generation of the material from which the ice of glaciers
is derived.   Mountain snow, the nutriment of glaciers,
is derived from aqueous vapour raised mainly from the
tropical ocean by the sun.   The solar fire is as necessary a factor in the process as our lamp in the experiment referred to a moment ago.   Nothing is easier
than to calculate the exact amount of heat expended
by the sun in the production of a glacier.   It would,
as I have elsewhere shown,[1] raise a quantity of cast
iron five times the weight of the glacier not only to a
white heat, but to its point of fusion.   If, as I have
already urged, instead of being filled with ice, the
valleys of the Alps were filled with white-hot metal, of
quintuple the mass of the present glaciers, it is the
heat, and not the cold, that would arrest our attention
and solicit our explanation.   The process of glacier
making is obviously one of distillation, in which the
fire of the sun, which generates the vapour, plays as
essential a part as the cold of the mountains which
condenses it.[2]

It was their ascription to glacier action that first

---

[1] 'Heat a Mode of Motion,' fifth edition, chap. vi.: Forms of
Water, §§ 55 and 56.

[2] In Lyell's excellent 'Principles of Geology,' the remark occurs
that 'several writers have fallen into the strange error of supposing
that the glacial period must have been one of higher mean temperature than usual.'   The really strange error was the forgetfulness
of the fact that without the heat the substance necessary to the
production of glaciers would be wanting.

gave the parallel roads of Glen Roy an interest in my eyes; and in 1867, with a view to self-instruction, I made a solitary pilgrimage to the place, and explored pretty thoroughly the roads of the principal glen. I traced the highest road to the col dividing Glen Roy from Glen Spey, and, thanks to the civility of an Ordnance surveyor, I was enabled to inspect some of the roads with a theodolite, and to satisfy myself regarding the common level of the shelves at opposite sides of the valley. As stated by Pennant, the width of the roads amounts sometimes to more than twenty yards; but near the head of Glen Roy the highest road ceases to have any width, for it runs along the face of a rock, the effect of the lapping of the water on the more friable portions of the rock being perfectly distinct to this hour. My knowledge of the region was, however, far from complete, and nine years had dimmed the memory even of the portion which had been thoroughly examined. Hence my desire to see the roads once more before venturing to talk to you about them. The Easter holidays of 1876 were to be devoted to this purpose; but at the last moment a telegram from Roy Bridge informed me that the roads were snowed up. Finding books and memories poor substitutes for the flavour of facts, I resolved subsequently to make another effort to see the roads. Accordingly last Thursday fortnight, after lecturing here, I packed up, and started (not this time alone) for the North. Next day at noon my wife and I found ourselves at Dalwhinnie, whence a drive of some five-and-thirty miles brought us to the excellent hostelry of Mr. Macintosh, at the mouth of Glen Roy.

We might have found the hills covered with mist, which would have wholly defeated us; but Nature was good-natured, and we had two successful working days among the hills. Guided by the excellent ordnance map of the region, on the Saturday morning we went up the glen, and on reaching the stream called Allt Bhreac Achaidh faced the hills to the west. At the watershed between Glen Roy and Glen Fintaig we bore northwards, struck the ridge above Glen Gluoy, came in view of its road, which we persistently followed as long as it continued visible. It is a feature of all the roads that they vanish before reaching the cols over which fell the waters of the lakes which formed them. One reason doubtless is that at their upper ends the lakes were shallow, and incompetent on this account to raise wavelets of any strength to act upon the mountain drift. A second reason is that they were land-locked in the higher portions and protected from the south-westerly winds, the stillness of their waters causing them to produce but a feeble impression upon the mountain sides. From Glen Gluoy we passed down Glen Turrit to Glen Roy, and through it homewards, thus accomplishing two or three and twenty miles of rough and honest work.

Next day we thoroughly explored Glen Glaster, following its two roads as far as they were visible. We reached the col discovered by Mr. Milne-Home, which stands at the level of the middle road of Glen Roy. Thence we crossed southwards over the mountain *Creag Dhubh,* and examined the erratic blocks upon its sides, and the ridges and mounds of moraine matter which cumber the lower flanks of the mountain. The observations of Mr. Jamieson upon this region, including the mouth of Glen Trieg, are in the highest degree interesting. We entered Glen Spean, and continued a search begun on the evening of our arrival at Roy Bridge— the search, namely, for glacier polishings and markings. We did not find them copious, but they are indubitable.

One of the proofs most convenient for reference, is a great rounded rock by the roadside, 1,000 yards east of the milestone marked three-quarters of a mile from Roy Bridge. Farther east other cases occur, and they leave no doubt upon the mind that Glen Spean was at one time filled by a great glacier. To the disciplined eye the aspect of the mountains is perfectly conclusive on this point; and in no position can the observer more readily and thoroughly convince himself of this than at the head of Glen Glaster. The dominant hills here are all intensely glaciated.

But the great collecting ground of the glaciers which dammed the glens and produced the parallel roads, were the mountains south and west of Glen Spean. The monarch of these is Ben Nevis, 4,370 feet high. The position of Ben Nevis and his colleagues, in reference to the vapour-laden winds of the Atlantic, is a point of the first importance. It is exactly similar to that of Carrantual and the Macgillicuddy Reeks in the south-west of Ireland. These mountains are, and were, the first to encounter the south-western Atlantic winds, and the precipitation, even at present, in the neighbourhood of Killarney, is enormous. The winds, robbed of their vapour, and charged with the heat set free by its precipitation, pursue their direction obliquely across Ireland; and the effect of the drying process may be understood by comparing the rainfall at Cahirciveen with that at Portarlington. As found by Dr. Lloyd, the ratio is as 59 to 21—fifty-nine inches annually at Cahirciveen to twenty-one at Portarlington. During the glacial epoch this vapour fell as snow, and the consequence was a system of glaciers which have left traces and evidences of the most impressive character in the region of the Killarney Lakes. I have referred in other places to the great glacier which, descending from the

Reeks, moved through the Black Valley, took possession of the lake-basins, and left its traces on every rock
and island emergent from the waters of the upper lake.
They are all conspicuously glaciated. Not in Switzerland itself do we find clearer traces of ancient glacier
action.

What the Macgillicuddy Reeks did in Ireland, Ben
Nevis and the adjacent mountains did, and continue to
do, in Scotland. We had an example of this on the
morning we quitted Roy Bridge. From the bridge
westward rain fell copiously, and the roads were wet;
but the precipitation ceased near Loch Laggan, whence
eastward the roads were dry. Measured by the gauge,
the rainfall at Fort William is 86 inches, while at
Laggan it is only 46 inches annually. The difference
between west and east is forcibly brought out by observations at the two ends of the Caledonian Canal. Fort
William at the south-western end has, as just stated,
86 inches, while Culloden, at its north-eastern end, has
only 24. To the researches of that able and accomplished meteorologist, Mr. Buchan, we are indebted for
these and other data of the most interesting and
valuable kind.

Adhering to the facts now presented to us, it is not
difficult to restore in idea the process by which the
glaciers of Lochaber were produced and the glens
dammed by ice. When the cold of the glacial epoch
began to invade the Scottish hills, the sun at the same
time acting with sufficient power upon the tropical
ocean, the vapours raised and drifted on to these
northern mountains were more and more converted
into snow. This slid down the slopes, and from every
valley, strath, and corry, south of Glen Spean, glaciers
were poured into that glen. The two great factors
here brought into play are the nutrition of the glaciers

by the frozen material above, and their consumption in the milder air below. For a period supply exceeded consumption, and the ice extended, filling Glen Spean to an ever-increasing height, and abutting against the mountains to the north of that glen. But why, it may be asked, should the valleys south of Glen Spean be receptacles of ice at a time when those north of it were receptacles of water? The answer is to be found in the position and the greater elevation of the mountains south of Glen Spean. They first received the loads of moisture carried by the Atlantic winds, and not until they had been in part dried, and also warmed by the liberation of their latent heat, did these winds touch the hills north of the Glen.

An instructive observation bearing upon this point is here to be noted. Had our visit been in the winter we should have found all the mountains covered; had it been in the summer we should have found the snow all gone. But happily it was at a season when the aspect of the mountains north and south of Glen Spean exhibited their relative powers as snow collectors. Scanning the former hills from many points of view, we were hardly able to detect a fleck of snow, while heavy swaths and patches loaded the latter. Were the glacial epoch to return, the relation indicated by this observation would cause Glen Spean to be filled with glaciers from the south, while the hills and valleys on the north, visited by warmer and drier winds, would remain comparatively free from ice. This flow from the south would be reinforced from the west, and as long as the supply was in excess of the consumption the glaciers would extend, the dams which closed the glens increasing in height. By-and-by supply and consumption becoming approximately equal, the height of the glacier barriers would remain constant. Then, if milder weather set in, consumption would be in excess, a lowering of the barriers and a retreat of the ice being the consequence. But for a long time the conflict between supply and consumption would continue, retarding indefinitely the disappearance of the barriers, and keeping the imprisoned lakes in the northern glens. But however slow its retreat, the ice in the long run would be forced to yield. The dam at the mouth of Glen Roy, which probably entered the glen sufficiently far to block up Glen Glaster, would gradually retreat. Glen Glaster and its col being opened, the subsidence of the lake eighty feet, from the level of the highest to that of the second parallel road, would follow as a consequence. I think this the most probable course of things, but it is also possible that Glen Glaster may have been blocked by a glacier from Glen Trieg. The ice dam continuing to retreat, at length permitted Glen Roy to connect itself with upper Glen Spean. A continuous lake then filled both glens, the level of which, as already explained, was determined by the col at Makul, above the head of Loch Laggan. The last to yield was the portion of the glacier which derived nutrition from Ben Nevis, and probably also from the mountains north and south of Loch Arkaig. But it at length yielded, and the waters in the glens resumed the courses which they pursue to-day.

For the removal of the ice barriers no cataclysm is to be invoked; the gradual melting of the dam would produce the entire series of phenomena. In sinking from col to col the water would flow over a gradually melting barrier, the surface of the imprisoned lake not remaining sufficiently long at any particular level to produce a shelf comparable to the parallel roads. By temporary halts in the process of melting due to atmospheric conditions or to the character of the dam itself, or through local softness in the drift, small pseudo-terraces would be formed, which, to the perplexity of some observers, are seen upon the flanks of the glens to-day.

In presence then of the fact that the barriers which stopped these glens to a height, it may be, of 1,500 feet above the bottom of Glen Spean, have dissolved and left not a wreck behind; in presence of the fact, insisted on by Professor Geikie, that barriers of detritus would undoubtedly have been able to maintain themselves had they ever been there; in presence of the fact that great glaciers once most certainly filled these valleys—that the whole region, as proved by Mr. Jamieson, is filled with the traces of their action; the theory which ascribes the parallel roads to lakes dammed by barriers of ice has, in my opinion, a degree of probability on its side which amounts to a practical demonstration of its truth.

Into the details of the terrace formation I do not enter. Mr. Darwin and Mr. Jamieson on the one side, and Sir John Lubbock on the other, deal with true causes. The terraces, no doubt, are due in part to the descending drift arrested by the water, and in part to the fretting of the wavelets, and the rearrangement of the stirred detritus, along the belts of contact of lake and hill. The descent of matter must have been frequent when the drift was unbound by the rootlets which hold it together now. In some cases, it may be remarked, the visibility of the roads is materially augmented by differences of vegetation. The grass upon the terraces is not always of the same character as that above and below them, while on heather-covered hills the absence of the dark shrub from the roads greatly enhances their conspicuousness.

The annexed sketch of a model (p. 225) will enable
the reader to grasp the essential features of the problem
and its solution.   Glen Gluoy and Glen Roy are lateral
valleys which open into Glen Spean.   Let us suppose
Glen Spean filled from v to w with ice of a uniform
elevation of 1,500 feet above the sea, the ice not
filling the upper part of that glen.   The ice would
thrust itself for some distance up the lateral valleys,
closing all their mouths.   The streams from the mountains right and left of Glen Gluoy would pour their
waters into that glen, forming a lake, the level of which
would be determined by the height of the col at ʌ, 1170
feet above the sea.   Over this col the water would flow
into Glen Roy.   But in Glen Roy it could not rise higher
than 1150 feet, the height of the col at ʙ, over which it
would flow into Glen Spey.

The water halting at these levels for a sufficient
time, would form the single road in Glen Gluoy and the
highest road in Glen Roy.   This state of things would
continue as long as the ice dam was sufficiently high to
dominate the cols at ʌ and ʙ; but when through change
of climate the gradually sinking dam reached, in succession, the levels of these cols, the water would then begin
to flow over the dam instead of over the cols.   Let us
suppose the wasting of the ice to continue until a connection was established between Glen Roy and Glen
Glaster, a common lake would then fill both these glens,
the level of which would be determined by that of the
col c, over which the water would pour for an indefinite
period into Glen Spean.   During this period the second
Glen Roy road and the highest road of Glen Glaster
would be formed.   The ice subsiding still further, a
connection would eventually be established between
Glen Roy, Glen Glaster, and the upper part of Glen
Spean.   A common lake would fill all three glens, the

FIG. 5.

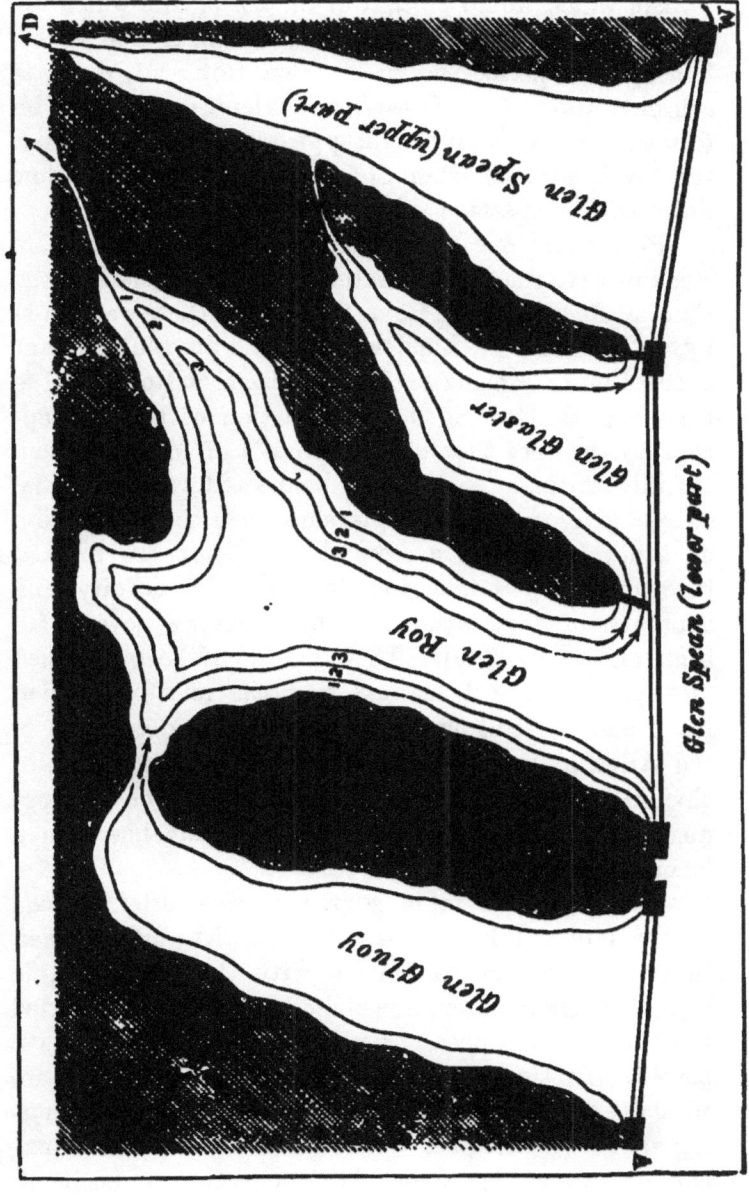

level of which would be that of the col D, over which for an indefinite period the lake would pour its water. During this period the lowest Glen Roy road, which is common also to Glen Glaster and Glen Spean, would be formed. Finally, on the disappearance of the ice from the lower part of Glen Spean the waters would flow down their respective valleys as they do to-day.

Reviewing our work, we find three considerable steps to have marked the solution of the problem of the Parallel Roads of Glen Roy. The first of these was taken by Sir Thomas Dick-Lauder, the second was the pregnant conception of Agassiz regarding glacier action, and the third was the testing and verification of this conception by the very thorough researches of Mr. Jamieson. No circumstance or incident connected with this discourse gives me greater pleasure than the recognition of the value of these researches. They are marked throughout by unflagging industry, by novelty and acuteness of observation, and by reasoning power of a high and varied kind. These pages had been returned ' for press ' when I learned that the relation of Ben Nevis and his colleagues to the vapour-laden winds of the Atlantic had not escaped Mr. Jamieson. To him obviously the exploration of Lochaber, and the development of the theory of the Parallel Roads, has been a labour of love.

Thus ends our rapid survey of this brief episode in the physical history of the Scottish hills,—brief, that is to say, in comparison with the immeasurable lapses of time through which, to produce its varied structure and appearances, our planet must have passed. In the survey of such a field two things are specially worthy to be taken into account—the widening of the intellectual horizon and the reaction of ex-panding knowledge upon the intellectual organ itself

MAP SHEWING THE PARALLEL ROADS OF GLEN ROY.

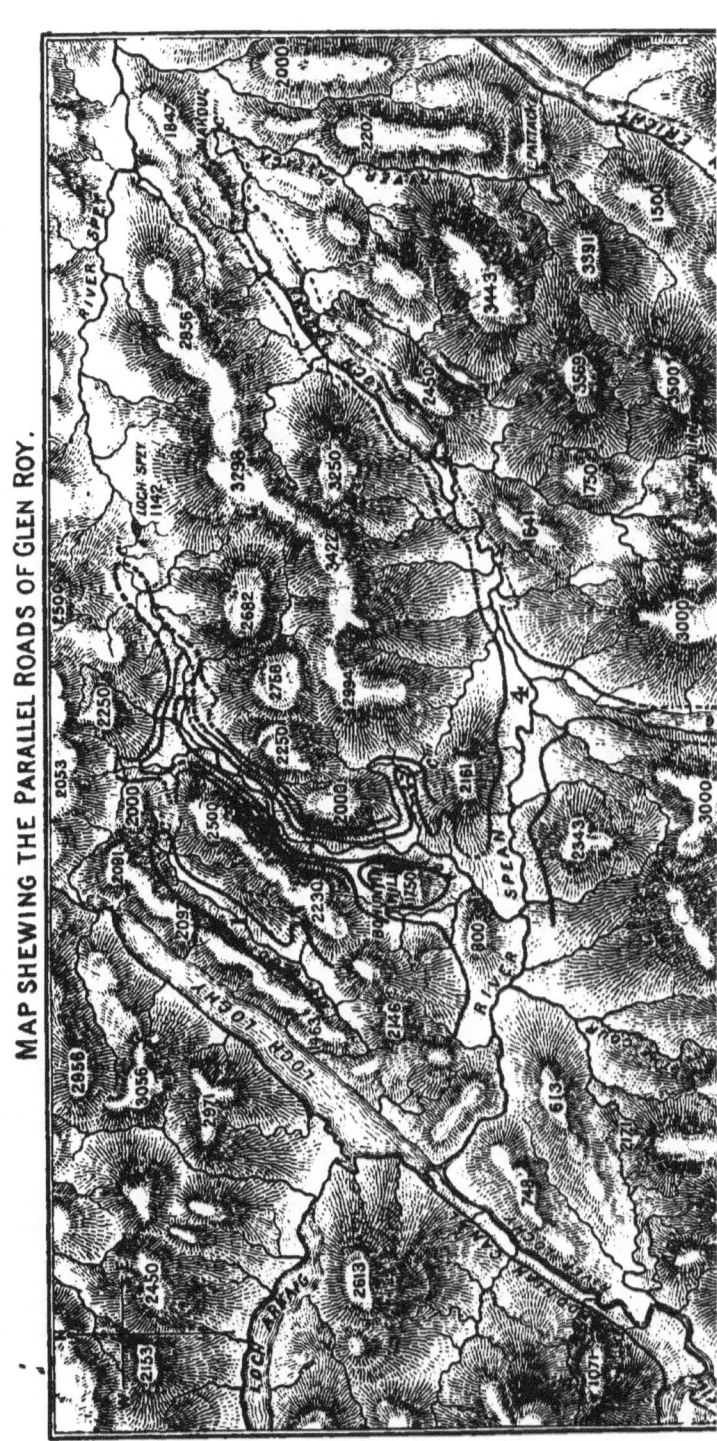

At first, as in the case of ancient glaciers, through sheer want of capacity, the mind refuses to take in revealed facts. But by degrees the steady contemplation of these facts so strengthens and expands the intellectual powers, that where truth once could not find an entrance it eventually finds a home.[1]

A map of the district, with the parallel roads shown in red, is annexed.

LITERATURE OF THE SUBJECT.

THOMAS PENNANT.—A Tour in Scotland. Vol. iii. 1776, p. 394.

JOHN MacCULLOCH.—On the Parallel Roads of Glen Roy. Geol. Soc. Trans. vol. iv. 1817, p. 314.

THOMAS LAUDER DICK (afterwards SIR THOMAS DICK-LAUDER, Bart.)—On the Parallel Roads of Lochaber. Edin. Roy. Soc. Trans. 1818, vol. ix. p. 1.

CHARLES DARWIN.—Observations on the Parallel Roads of Glen Roy, and of the other parts of Lochaber in Scotland, with an attempt to prove that they are of marine origin. Phil. Trans. 1839, vol. cxxix. p. 39.

SIR CHARLES LYELL.—Elements of Geology. Second edition, 1841.

LOUIS AGASSIZ.—The Glacial Theory and its Recent Progress— Parallel Terraces. Edin. New Phil. Journal, 1842, vol. xxxiii. p. 236.

---

[1] The formation, connection, successive subsidence, and final disappearance of the glacial lakes of Lochaber were illustrated in the discourse here reported by the model just described, constructed under the supervision of my assistant, Mr. John Cottrell. Glen Gluoy with its lake and road and the cataract over its col; Glen Roy and its three roads with their respective cataracts at the head of Glen Spey, Glen Glaster, and Glen Spean, were all represented. The successive shiftings of the barriers, which were formed of plate glass, brought each successive lake and its corresponding road into view, while the entire removal of the barriers caused the streams to flow down the glens of the model as they flow down the real glens of to-day.

DAVID MILNE (afterwards DAVID MILNE-HOME).—On the Parallel Roads of Lochaber; with Remarks on the Change of Relative Levels of Sea and Land in Scotland, and on the Detrital Deposits in that Country. Edin. Roy. Soc. Trans. 1847, vol. xvi. p. 395.

ROBERT CHAMBERS.—Ancient Sea Margins. Edinburgh, 1848.

H. D. ROGERS.—On the Parallel Roads of Glen Roy. Royal Inst. Proceedings, 1861, vol. iii. p. 341.

THOMAS F. JAMIESON.—On the Parallel Roads of Glen Roy, and their Place in the History of the Glacial Period. Quart. Journal Geol. Soc. 1863, vol. xix. p. 235.

SIR CHARLES LYELL.—Antiquity of Man. 1863, p. 253.

REV. R. B. WATSON.—On the Marine Origin of the Parallel Roads of Glen Roy. Quart. Journ. Geol. Soc. 1865, vol. xxii. p. 9.

SIR JOHN LUBBOCK.—On the Parallel Roads of Glen Roy. Quart. Journ. Geol. Soc. 1867, vol. xxiv. p. 83.

CHARLES BABBAGE.—Observations on the Parallel Roads of Glen Roy. Quart. Journ. Geol. Soc. 1868, vol. xxiv. p. 273.

JAMES NICOL.—On the Origin of the Parallel Roads of Glen Roy. 1869. Geol. Soc. Journal, vol. xxv. p. 282.

JAMES NICOL.—How the Parallel Roads of Glen Roy were formed. 1872. Geol. Soc. Journal, vol. xxviii. p. 237.

MAJOR-GENERAL SIR HENRY JAMES, R. E..—Notes on the Parallel Roads of Lochaber. 4to. 1874.

IX.

ALPINE SCULPTURE.

1864.

TO account for the conformation of the Alps, two hypotheses have been advanced, which may be respectively named the hypothesis of *fracture* and the hypothesis of *erosion*. The former assumes that the forces by which the mountains were elevated produced fissures in the earth's crust, and that the valleys of the Alps are the tracks of these fissures ; while the latter maintains that the valleys have been cut out by the action of ice and water, the mountains themselves being the residual forms of this grand sculpture. I had heard the Via Mala cited as a conspicuous illustration of the fissure theory—the profound chasm thus named, and through which the Hinter-Rhein now flows, could, it was alleged, be nothing else than a crack in the earth's crust. To the Via Mala I therefore went in 1864 to instruct myself upon the point in question.

The gorge commences about a quarter of an hour above Tusis ; and, on entering it, the first impression certainly is that it must be a fissure. This conclusion in my case was modified as I advanced. Some distance up the gorge I found upon the slopes to my right quantities of rolled stones, evidently rounded by water-action. Still further up, and just before reaching the first bridge which spans the chasm, I found more rolled

stones, associated with sand and gravel. Through this
mass of detritus, fortunately, a vertical cutting had
been made, which exhibited a section showing perfect
stratification. There was no agency in the place to roll
these stones, and to deposit these alternating layers of
sand and pebbles, but the river which now rushes some
hundreds of feet below them. At one period of the Via
Mala's history the river must have run at this high
level. Other evidences of water-action soon revealed
themselves. From the parapet of the first bridge I
could see the solid rock 200 feet above the bed of the
river scooped and eroded.

It is stated in the guide-books that the river, which
usually runs along the bottom of the gorge, has been
known almost to fill it during violent thunder-storms; and
it may be urged that the marks of erosion which the sides
of the chasm exhibit are due to those occasional floods.
In reply to this, it may be stated that even the exit-t-
ence of such floods is not well authenticated, and that
if the supposition were true, it would be an additional
argument in favour of the cutting power of the river.
For if floods operating at rare intervals could thus
erode the rock, the same agency, acting without ceasing
upon the river's bed, must certainly be competent to
excavate it.

I proceeded upwards, and from a point near another
bridge (which of them I did not note) had a fine view
of a portion of the gorge. The river here runs at the
bottom of a cleft of profound depth, but so narrow that
it might be leaped across. That this cleft must be a
crack is the impression first produced; but a brief inspection suffices to prove that it has been cut by the
river. From top to bottom we have the unmistakable
marks of erosion. This cleft was best seen on looking
downwards from a point near the bridge; but looking

upwards from the bridge itself, the evidence of aqueous erosion was equally convincing.

The character of the erosion depends upon the rock as well as upon the river. The action of water upon some rocks is almost purely mechanical; they are simply ground away or detached in sensible masses. Water, however, in passing over limestone, charges it self with carbonate of lime without damage to its trans parency; the rock is *dissolved* in the water; and the gorges cut by water in such rocks often re emble those cut in the ice of glaciers by glacier streams. To the solubility of limestone is probably to be ascribed the fantastic forms which peaks of this rock usually assume, and also the grottos and caverns which interpenetrate limestone formations. A rock capable of being thus dissolved will expose a smooth surface after the water has quitted it; and in the case of the Via Mala it is the polish of the surfaces and the curved hollows scooped in the sides of the gorge, which assure us that the chasm has been the work of the river.

About four miles from Tusis, and not far from the little village of Zillis, the Via Mala opens into a plain bounded by high terraces. It occurred to me the moment I saw it that the plain had been the bed of an ancient lake; and a farmer, who was my temporary companion, immediately informed me that such was the tradition of the neighbourhood. This man conversed with intelligence, and as I drew his attention to the rolled stones, which rest not only above the river, but above the road, and inferred that the river must once have been there to have rolled those stones, he saw the force of the evidence perfectly. In fact, in former times, and subsequent to the retreat of the great glaciers, a rocky barrier crossed the valley at this place, damming the river which came from the mountains

16

higher up. A lake was thus formed which poured its
waters over the barrier. Two actions were here at work,
both tending to obliterate the lake—the raising of its
bed by the deposition of detritus, and the cutting of its
dam by the river. In process of time the cut deepened
into the Via Mala; the lake was drained, and the river
now flows in a definite channel through the plain which
its waters once totally covered.

From Tusis I crossed to Tiefenkasten by the Schien
Pass, and thence over the Julier Pass to Pontresina.
There are three or four ancient lake-beds between Tiefenkasten and the summit of the Julier. They are all of
the same type—a more or less broad and level valley-
bottom, with a barrier in front through which the
river has cut a passage, the drainage of the lake being
the consequence. These lakes were sometimes dammed
by barriers of rock, sometimes by the moraines of
ancient glaciers.

An example of this latter kind occurs in the Rosegg
valley, about twenty minutes below the end of the
Rosegg glacier, and about an hour from Pontresina.
The valley here is crossed by a pine-covered moraine
of the noblest dimensions; in the neighbourhood of
London it might be called a mountain. That it is a
moraine, the inspection of it from a point on the Surlei
slopes above it will convince any person possessing an
educated eye. Where, moreover, the interior of the
mound is exposed, it exhibits moraine-matter—detritus
pulverised by the ice, with boulders entangled in it.
It stretched quite across the valley, and at one time
dammed the river up. But now the barrier is cut
through, the stream having about one-fourth of the
moraine to its right, and the remaining three-fourths
to its left. Other moraines of a more resisting character hold their ground as barriers to the present day.

In the Val di Campo, for example, about three-quarters of an hour from Pisciadello, there is a moraine composed of large boulders, which interrupt the course of a river and compel the water to fall over them in cascades. They have in great part resisted its action since the retreat of the ancient glacier which formed the moraine. Behind the moraine is a lake-bed, now converted into a level meadow, which rests on a deep layer of mould.

At Pontresina a very fine and instructive gorge is to be seen. The river from the Morteratsch glacier rushes through a deep and narrow chasm which is spanned at one place by a stone bridge. The rock is not of a character to preserve smooth polishing; but the larger features of water-action are perfectly evident from top to bottom. Those features are in part visible from the bridge, but still better from a point a little distance from the bridge in the direction of the upper village of Pontresina. The hollowing out of the rock by the eddies of the water is here quite manifest. A few minutes' walk upwards brings us to the end of the gorge ; and behind it we have the usual indications of an ancient lake, and terraces of distinct water origin. From this position indeed the genesis of the gorge is clearly revealed. After the retreat of the ancient glacier, a transverse ridge of comparatively resisting material crossed the valley at this place. Over the lowest part of this ridge the river flowed, rushing steeply down to join at the bottom of the slope the stream which issued from the Rosegg glacier. On this incline the water became a powerful eroding agent, and finally cut the channel to its present depth.

Geological writers of reputation assume at this place the existence of a fissure, the ' washing out ' of which resulted in the formation of the gorge. Now no examination of the bed of the river ever proved the existence of this fissure ; and it is certain that water, particularly when charged with solid matter in suspension, can cut a channel through unfissured rock.   Cases of deep cutting can be pointed out where the clean bed of the stream is exposed, the rock which forms the floor of the river not exhibiting a trace of fissure.   An example of this kind on a small scale occurs near the Bernina Gasthaus, about two hours from Pontresina.   A little way below the junction of the two streams from the Bernina Pass and the Heuthal the river flows through a channel cut by itself, and 20 or 30 feet in depth.   At some places the river-bed is covered with rolled stones ; at other places it is bare, but shows no trace of fissure.   The abstract power of water, if I may use the term, to cut through rock is demonstrated by such instances.   But if water be competent to form a gorge without the aid of a fissure, why assume the existence of such fissures in cases like that at Pontresina ?   It seems far more philosophical to accept the simple and impressive history written on the walls of those gorges by the agent which produced them.

Numerous cases might be pointed out, varying in magnitude, but all identical in kind, of barriers which crossed valleys and formed lakes having been cut through by rivers, narrow gorges being the consequence.   One of the most famous examples of this kind is the Finster-aarschlucht in the valley of Hasli.   Here the ridge called the Kirchet seems split across, and the river Aar rushes through the fissure.   Behind the barrier we have the meadows and pastures of Imhof resting on the sediment of an ancient lake.   Were this an isolated case, one might with an apparent show of reason conclude that the Finsteraarschlucht was produced by an earthquake, as some suppose it to have been ; but when we find it to be a single sample of actions which are frequent in the Alps

—when probably a hundred cases of the same kind, though different in magnitude, can be pointed out—it seems quite unphilosophical to assume that in each particular case an earthquake was at hand to form a channel for the river. As in the case of the barrier at Pontresina, the Kirchet, after the retreat of the Aar glacier, dammed the waters flowing from it, thus forming a lake, on the bed of which now stands the village of Imhof. Over this barrier the Aar tumbled towards Meyringen, cutting, as the centuries passed, its bed ever deeper, until finally it became deep enough to drain the lake, leaving in its place the alluvial plain, through which the river now flows in a definite channel.

In 1866 I subjected the Finsteraarschlucht to a close examination. The earthquake theory already adverted to was then prevalent regarding it, and I wished to see whether any evidences existed of aqueous erosion. Near the summit of the Kirchet is a signboard inviting the traveller to visit the *Aarenschlucht*, a narrow lateral gorge which runs down to the very bottom of the principal one. The aspect of this smaller chasm from bottom to top proves to demonstration that water had in former ages been there at work. It is scooped, rounded, and polished, so as to render palpable to the most careless eye that it is a gorge of erosion. But it was regarding the sides of the great chasm that instruction was needed, and from its edge nothing to satisfy me could be seen. I therefore stripped and waded into the river until a point was reached which commanded an excellent view of both sides of the gorge. The water was cutting cold, but I was repaid. Below me on the left-hand side was a jutting cliff which bore the thrust of the river and caused the Aar to swerve from its direct course. From top to bottom this cliff was polished, rounded, and scooped. There was no room for doubt. The river which now runs so deeply down had once been above. It has been the delver of its own channel through the barrier of the Kirchet.

But the broad view taken by the advocates of the fracture theory is, that the valleys themselves follow the tracks of primeval fissures produced by the upheaval of the land, the cracks across the barriers referred to being in reality portions of the great cracks which formed the valleys. Such an argument, however, would virtually concede the theory of erosion as applied to the valleys of the Alps. The narrow gorges, often not more than twenty or thirty feet across, sometimes even narrower, frequently occur at the bottom of broad valleys. Such fissures might enter into the list of accidents which gave direction to the real erosive agents which scooped the valley out; but the formation of the valley, as it now exists, could no more be ascribed to such cracks than the motion of a railway train could be ascribed to the finger of the engineer which turns on the steam.

These deep gorges occur, I believe, for the most part in limestone strata; and the effects which the merest driblet of water can produce on limestone are quite astonishing. It is not uncommon to meet chasms of considerable depth produced by small streams the beds of which are dry for a large portion of the year. Right and left of the larger gorges such secondary chasms are often found. The idea of *time* must, I think, be more and more included in our reasonings on these phenomena. Happily, the marks which the rivers have, in most cases, left behind them, and which refer, geologically considered, to actions of yesterday, give us ground and courage to conceive what may be effected in geologic periods. Thus the modern portion of the Via Mala throws light upon the whole. Near Bergün, in the valley of the Albula, there is also a little Via Mala, which is not less significant than the great one. The river flows here through a profound limestone gorge, and to the very edges of the gorge we have the evidences of erosion. But the most striking illustration of water-action upon limestone rock that I have ever seen is the gorge at Pfäffers. Here the traveller passes along the side of the chasm midway between top and bottom. Whichever way he looks, backwards or forwards, upwards or downwards, towards the sky or towards the river, he meets everywhere the irresistible and impressive evidence that this wonderful fissure has been sawn through the mountain by the waters of the Tamina.

I have thus far confined myself to the consideration of the gorges formed by the cutting through of the rock-barriers which frequently cross the valleys of the Alps; as far as they have been examined by me they are the work of erosion. But the larger question still remains, To what action are we to ascribe the formation of the valleys themselves? This question includes that of the formation of the mountain-ridges, for were the valleys wholly filled, the ridges would disappear. Possibly no answer can be given to this question which is not beset with more or less of difficulty. Special localities might be found which would seem to contradict every solution which refers the conformation of the Alps to the operation of a single cause.

Still the Alps present features of a character sufficiently definite to bring the question of their origin within the sphere of close reasoning. That they were in whole or in part once beneath the sea will not be disputed; for they are in great part composed of sedimentary rocks which required a sea to form them. Their present elevation above the sea is due to one of those local changes in the shape of the earth which have been of frequent occurrence throughout geologic time, in some cases depressing the land, and in others causing the sea-bottom to protrude beyond its surface. Considering the inelastic character of its materials, the protuberance of the Alps could hardly have been pushed out without dislocation and fracture; and this conclusion gains in probability when we consider the foldings, contortions, and even reversals in position of the strata in many parts of the Alps. Such changes in the position of beds which were once horizontal could not have been effected without dislocation. Fissures would be produced by these changes; and such fissures, the advocates of the fracture theory contend, mark the positions of the valleys of the Alps.

Imagination is necessary to the man of science, and we could not reason on our present subject without the power of presenting mentally a picture of the earth's crust cracked and fissured by the forces which produced its upheaval. Imagination, however, must be strictly checked by reason and by observation. That fractures occurred cannot, I think, be doubted, but that the valleys of the Alps are thus formed is a conclusion not at all involved in the admission of dislocations. I never met with a precise statement of the manner in which the advocates of the fissure theory suppose the forces to have acted—whether they assume a general elevation of the region, or a local elevation of distinct ridges; or whether they assume local subsidences after a general elevation, or whether they would superpose upon the general upheaval minor and local upheavals.

In the absence of any distinct statement, I will assume the elevation to be general—that a swelling out of the earth's crust occurred here, sufficient to place the most prominent portions of the protuberance three miles above the sea-level. To fix the ideas, let us consider a circular portion of the crust, say one hundred miles in diameter, and let us suppose, in the first instance, the circumference of this circle to remain fixed, and that the elevation was confined to the space within it. The upheaval would throw the crust into a state of strain; and, if it were inflexible, the strain must be relieved by fracture. Crevasses would thus intersect the crust. Let us now enquire what proportion the area of these open fissures is likely to bear to the area of the unfissured crust. An approximate answer is all that is here required; for the problem is of such a character as to render minute precision unnecessary.

No one, I think, would affirm that the area of the fissures would be one-hundredth the area of the land. For let us consider the strain upon a single line drawn over the summit of the protuberance from a point on its rim to a point opposite. Regarding the protuberance as a spherical swelling, the length of the arc corresponding to a chord of 100 miles and a versed sine of 3 miles is 100·24 miles; consequently the surface to reach its new position must stretch 0·24 of a mile, or be broken. A fissure or a number of cracks with this total width would relieve the strain; that is to say, the sum of the widths of all the cracks over the length of 100 miles would be 420 yards. If, instead of comparing the width of the fissures with the length of the lines of tension, we compared their areas with the area of the unfissured land, we should of course find the proportion much less. These considerations will help the imagination to realise what a small ratio the area of the open fissures must bear to the unfissured crust. They enable us to say, for example, that to assume the area of the fissures to be one-tenth of the area of the land would be quite absurd, while that the area of the fissures could be one-half or more than one-half that of the land would be in a proportionate degree unthinkable. If we suppose the elevation to be due to the shrinking or subsidence of the land all round our assumed circle, we arrive equally at the conclusion that the area of the open fissures would be altogether insignificant as compared with that of the unfissured crust.

To those who have seen them from a commanding elevation, it is needless to say that the Alps themselves bear no sort of resemblance to the picture which this theory presents to us. Instead of deep cracks with approximately vertical walls, we have ridges running into peaks, and gradually sloping to form valleys. Instead of a fissured crust, we have a state of things closely resembling the surface of the ocean when agitated by a storm. The valleys, instead of being much narrower than the ridges, occupy the greater space. A plaster cast of the Alps turned upside down, so as to invert the elevations and depressions, would exhibit blunter and broader mountains, with narrower valleys between them, than the present ones. The valleys that exist cannot, I think, with any correctness of language be called fissures. It may be urged that they originated in fissures: but even this is unproved, and, were it proved, the fissures would still play the subordinate part of giving direction to the agents which are to be regarded as the real sculptors of the Alps.

The fracture theory, then, if it regards the elevation of the Alps as due to the operation of a force acting throughout the entire region, is, in my opinion, utterly incompetent to account for the conformation of the country. If, on the other hand, we are compelled to resort to local disturbances, the manipulation of the earth's crust necessary to obtain the valleys and the mountains will, I imagine, bring the difficulties of the theory into very strong relief. Indeed an examination

of the region from many of the more accessible eminences—from the Galenstock, the Grauhaupt, the Pitz
Languard, the Monte Confinale—or, better still, from
Mont Blanc, Monte Rosa, the Jungfrau, the Finsteraar-
horn, the Weisshorn, or the Matterhorn, where local
peculiarities are toned down, and the operations of the
powers which really made this region what it is are
alone brought into prominence—must, I imagine, convince every physical geologist of the inability of any
fracture theory to account for the present conformation
of the Alps.

A correct model of the mountains, with an un-
exaggerated vertical scale, produces the same effect
upon the mind as the prospect from one of the highest
peaks. We are apt to be influenced by local phenomena
which, though insignificant in view of the general
question of Alpine conformation, are, with reference to
our customary standards, vast and impressive. In a
true model those local peculiarities disappear; for on
the scale of a model they are too small to be visible;
while the essential facts and forms are presented to the
undistracted attention.

A minute analysis of the phenomena strengthens
the conviction which the general aspect of the Alps
fixes in the mind. We find, for example, numerous
valleys which the most ardent plutonist would not
think of ascribing to any other agency than erosion.
That such is their genesis and history is as certain as
that erosion produced the Chines in the Isle of Wight.
From these indubitable cases of erosion—commencing,
if necessary, with the small ravines which run down
the flanks of the ridges, with their little working
navigators at their bottoms—we can proceed, by almost
insensible gradations, to the largest valleys of the
Alps; and it would perplex the plutonist to fix upon

the point at which fracture begins to play a material part.

In ascending one of the larger valleys, we enter it where it is wide and where the eminences are gentle on either side. The flanking mountains become higher and more abrupt as we ascend, and at length we reach a place where the depth of the valley is a maximum. Continuing our walk upwards, we find ourselves flanked by gentler slopes, and finally emerge from the valley and reach the summit of an open col, or depression in the chain of mountains. This is the common character of the large valleys. Crossing the col, we descend along the opposite slope of the chain, and through the same series of appearances in the reverse order. If the valleys on both sides of the col were produced by fissures, what prevents the fissure from prolonging itself across the col? The case here cited is representative; and I am not acquainted with a single instance in the Alps where the chain has been cracked in the manner indicated. The cols are simply depressions; in many of which the unfissured rock can be traced from side to side.

The typical instance just sketched follows as a natural consequence from the theory of erosion. Before either ice or water can exert great power as an erosive agent, it must collect in sufficient mass. On the higher slopes and plateaus—in the region of cols—the power is not fully developed; but lower down tributaries unite, erosion is carried on with increased vigour, and the excavation gradually reaches a maximum. Lower still the elevations diminish and the slopes become more gentle; the cutting power gradually relaxes, until finally the eroding agent quits the mountains altogether, and the grand effects which it produced in the earlier portions of its course entirely disappear.

I have hitherto confined myself to the consideration of the broad question of the erosion theory as compared with the fracture theory; and all that I have been able to observe and think with reference to the subject leads me to adopt the former. Under the term erosion I include the action of water, of ice, and of the atmosphere, including frost and rain. Water and ice, however, are the principal agents, and which of these two has produced the greatest effect it is perhaps impossible to say. Two years ago I wrote a brief note 'On the Conformation of the Alps,'[1] in which I ascribed the paramount influence to glaciers. The facts on which that opinion was founded are, I think, unassailable; but whether the conclusion then announced fairly follows from the facts is, I confess, an open question.

The arguments which have been thus far urged against the conclusion are not convincing. Indeed, the idea of glacier erosion appears so daring to some minds that its boldness alone is deemed its sufficient refutation. It is, however, to be remembered that a precisely similar position was taken up by many excellent workers when the question of ancient glacier extension was first mooted. The idea was considered too hardy to be entertained; and the evidences of glacial action were sought to be explained by reference to almost any process rather than the true one. Let those who so wisely took the side of ' boldness ' in that discussion beware lest they place themselves, with reference to the question of glacier erosion, in the position formerly occupied by their opponents.

Looking at the little glaciers of the present day— mere pigmies as compared to the giants of the glacial epoch—we find that from every one of them issues a

[1] Phil. Mag. vol. xxiv. p. 169.

river more or less voluminous, charged with the matter which the ice has rubbed from the rocks. Where the rocks are soft, the amount of this finely pulverised matter suspended in the water is very great. The water, for example, of the river which flows from Santa Catarina to Bormio is thick with it. The Rhine is charged with this matter, and by it has so silted up the Lake of Constance as to abolish it for a large fraction of its length. The Rhone is charged with it, and tens of thousands of acres of cultivable land are formed by the silt above the Lake of Geneva.

In the case of every glacier we have two agents at work—the ice exerting a crushing force on every point of its bed which bears its weight, and either rasping this point into powder or tearing it bodily from the rock to which it belongs; while the water which every-where circulates upon the bed of the glacier continually washes the detritus away and leaves the rock clean for further abrasion. Confining the action of glaciers to the simple rubbing away of the rocks, and allowing them sufficient time to act, it is not a matter of opinion, but a physical certainty, that they will scoop out valleys. But the glacier does more than abrade. Rocks are not homogeneous; they are intersected by joints and places of weakness, which divide them into virtually detached masses. A glacier is undoubtedly competent to root such masses bodily away. Indeed the mere à priori consideration of the subject proves the competence of a glacier to deepen its bed. Taking the case of a glacier 1,000 feet deep (and some of the older ones were probably three times this depth), and allowing 40 feet of ice to an atmosphere, we find that on every square inch of its bed such a glacier presses with a weight of 375 lbs., and on every square yard of its bed with a weight of 486,000 lbs. With a *vertical* pressure of this amount the glacier is urged down its valley by the pressure from behind. We can hardly, I think, deny to such a tool a power of excavation.

The retardation of a glacier by its bed has been referred to as proving its impotence as an erosive agent; but this very retardation is in some measure an expression of the magnitude of the erosive energy. Either the bed must give way, or the ice must slide over itself. We get indeed some idea of the crushing pressure which the moving glacier exercises against its bed from the fact that the resistance, and the effort to overcome it, are such as to make the upper layers of a glacier move bodily over the lower ones—a portion only of the total motion being due to the progress of the entire mass of the glacier down its valley.

The sudden bend in the valley of the Rhone at Martigny has also been regarded as conclusive evidence against the theory of erosion. 'Why,' it has been asked, 'did not the glacier of the Rhone go straight forward instead of making this awkward bend?' But if the valley be a crack, why did the crack make this bend? The crack, I submit, had at least as much reason to prolong itself in a straight line as the glacier had. A statement of Sir John Herschel with reference to another matter is perfectly applicable here: 'A crack once produced has a tendency to run—for this plain reason, that at its momentary limit, at the point at which it has just arrived, the divellent force on the molecules there situated is counteracted only by half of the cohesive force which acted when there was no crack, viz. the cohesion of the uncracked portion alone' ('Proc. Roy. Soc.' vol. xii. p. 678). To account, then, for the bend, the adherent of the fracture theory must assume the existence of some accident which turned the crack at right angles to itself; and he surely will permit the adherent of the erosion theory to make a similar assumption.

The influence of small accidents on the direction of rivers is beautifully illustrated in glacier streams, which are made to cut either straight or sinuous channels by causes apparently of the most trivial character. In his interesting paper ' On the Lakes of Switzerland,' M. Studer also refers to the bend of the Rhine at Sargans in proof that the river must there follow a pre-existing fissure. I made a special expedition to the place in 1864 ; and though it was plain that M. Studer had good grounds for the selection of this spot, I was unable to arrive at his conclusion as to the necessity of a fissure.

Again, in the interesting volume recently published by the Swiss Alpine Club, M. Desor informs us that the Swiss naturalists who met last year at Samaden visited the end of the Morteratsch glacier, and there convinced themselves that a glacier had no tendency whatever to imbed itself in the soil. I scarcely think that the question of glacier erosion, as applied either to lakes cr valleys, is to be disposed of so easily. Let me record here my experience of the Morteratsch glacier. I took with me in 1864 a theodolite to Pontresina, and while there had to congratulate myself on the aid of my friend Mr. Hirst, who in 1857 did such good service upon the Mer de Glace and its tributaries. We set out three lines across the Morteratsch glacier, one of which crossed the ice-stream near the well-known hut of the painter Georgei, while the two others were staked out, the one above the hut and the other below it. Calling the highest line A, the line which crossed the glacier at the hut B, and the lowest line C, the following are the mean hourly motions of the three lines, deduced from observations which extended over several days. On each line eleven stakes were fixed, which are designated by the figures 1, 2, 3, &c. in the Tables.

Morteratsch Glacier, Line A.

No. of Stake.	Hourly Motion.
1	0·35 inch.
2	0·49 „
3	0·53 „
4	0·54 „
5	0·56 „
6	0·54 „
7	0·52 „
8	0·49 „
9	0·40 „
10	0·29 „
11	0·20 „

As in all other measurements of this kind, the retarding influence of the sides of the glacier is manifest: the centre moves with the greatest velocity.

Morteratsch Glacier, Line B.

No. of Stake.	Hourly Motion.
1	0·05 inch.
2	0·14 „
3	0·24 „
4	0·32 „
5	0·41 „
6	0·44 „
7	0·44 „
8	0·45 „
9	0·43 „
10	0·44 „
11	0·44 „

The first stake of this line was quite close to the edge of the glacier, and the ice was thin at the place, hence its slow motion. Crevasses prevented us from carrying the line sufficiently far across to render the retardation of the further side of the glacier fully evident.

17

Morteratsch Glacier, Line C.

No. of Stake						Hourly Motion.
1	.	.	.	.	.	. 0·05 inch.
2	.	.	.	.	.	. 0·09 „
3	.	.	.	.	.	. 0·18 „
4	.	.	.	.	.	. 0·20 „
5	.	.	.	.	.	. 0·25 „
6	.	.	.	.	.	. 0·27 „
7	.	.	.	.	.	. 0·27 „
8	.	.	.	.	.	. 0·30 „
9	.	.	.	.	.	. 0·21 „
10	.	.	.	.	.	. 0·20 „
11	.	.	.	.	.	. 0·16 „

Comparing the three lines together, it will be observed that the velocity diminishes as we descend the glacier. In 100 hours the maximum motion of the three lines respectively is as follows :

Maximum Motion in 100 *hours.*

Line A	.	.	.	.	. 56 inches
„ B	.	.	.	.	. 45 „
„ C	.	.	.	.	. 30 „

This deportment explains an appearance which must strike every observer who looks upon the Morteratsch from the Piz Languard, or from the new Bernina Road. A medial moraine runs along the glacier, commencing as a narrow streak, but towards the end the moraine extending in width, until finally it quite covers the terminal portion of the glacier. The cause of this is revealed by the foregoing measurements, which prove that a stone on the moraine where it is crossed by the line A approaches a second stone on the moraine where it is crossed by the line C with a velocity of twenty-six inches per one hundred hours. The moraine is in a state of longitudinal compression. Its materials are more and more squeezed together, and they must consequently move laterally and render the moraine at the terminal portion of the glacier wider than above.

The motion of the Morteratsch glacier, then, diminishes as we descend. The maximum motion of the third line is thirty inches in one hundred hours, or seven inches a day—a very slow motion; and had we run a line nearer to the end of the glacier, the motion would have been slower still. At the end itself it is nearly insensible.[1] Now I submit that this is not the place to seek for the scooping power of a glacier. The opinion appears to be prevalent that it is the snout of a glacier that must act the part of ploughshare; and it is certainly an erroneous opinion. The scooping power will exert itself most where the weight and the motion are greatest. A glacier's snout often rests upon matter which has been scooped from the glacier's bed higher up. I therefore do not think that the inspection of what the end of a glacier does or does not accomplish can decide this question.

The snout of a glacier is potent to remove anything against which it can fairly abut; and this power, notwithstanding the slowness of the motion, manifests itself at the end of the Morteratsch glacier. A hillock, bearing pine-trees, was in front of the glacier when Mr. Hirst and myself inspected its end; and this hillock is being bodily removed by the thrust of the ice. Several of the trees are overturned; and in a few years, if the glacier continues its reputed advance, the mound will certainly be ploughed away.

The question of Alpine conformation stands, I think, thus: We have, in the first place, great valleys,

---

[1] The snout of the Aletsch Glacier has a diurnal motion of less than two inches, while a mile or so above the snout the velocity is eighteen inches. The spreading out of the moraine is here very striking.

such as those of the Rhine and the Rhone, which we might conveniently call valleys of the first order. The mountains which flank these main valleys are also cut by lateral valleys running into the main ones, and which may be called valleys of the second order. When these latter are examined, smaller valleys are found running into them, which may be called valleys of the third order. Smaller ravines and depressions, again, join the latter, which may be called valleys of the fourth order, and so on until we reach streaks and cuttings so minute as not to merit the name of valleys at all. At the bottom of every valley we have a stream, diminishing in magnitude as the order of the valley ascends, carving the earth and carrying its materials to lower levels. We find that the larger valleys have been filled for untold ages by glaciers of enormous dimensions, always moving, grinding down and tearing away the rocks over which they passed. We have, moreover, on the plains at the feet of the mountains, and in enormous quantities, the very matter derived from the sculpture of the mountains themselves.

The plains of Italy and Switzerland are cumbered by the *débris* of the Alps. The lower, wider, and more level valleys are also filled to unknown depths with the materials derived from the higher ones. In the vast quantities of moraine-matter which cumber many even of the higher valleys we have also suggestions as to the magnitude of the erosion which has taken place. This moraine-matter, moreover, can only in small part have been derived from the falling of rocks *upon* the ancient glacier; it is in great part derived from the grinding and the ploughing-out of the glacier itself. This accounts for the magnitude of many of the ancient moraines, which date from a period when almost all the mountains were covered with ice and snow, and when, consequently, the quantity of moraine-matter derived from the naked crests cannot have been considerable.

The erosion theory ascribes the formation of Alpine valleys to the agencies here briefly referred to. It invokes nothing but true causes. Its artificers are still there, though, it may be, in diminished strength; and if they are granted sufficient time, it is demonstrable that they are competent to produce the effects ascribed to them. And what does the fracture theory offer in comparison? From no possible application of this theory, pure and simple, can we obtain the slopes and forms of the mountains. Erosion must in the long run be invoked, and its power therefore conceded. The fracture theory infers from the disturbances of the Alps the existence of fissures; and this is a probable inference. But that they were of a magnitude sufficient to produce the conformation of the Alps, and that they followed, as the Alpine valleys do, the lines of natural drainage of the country, are assumptions which do not appear to me to be justified either by reason or by observation.

There is a grandeur in the secular integration of small effects implied by the theory of erosion almost superior to that involved in the idea of a cataclysm. Think of the ages which must have been consumed in the execution of this colossal sculpture. The question may, of course, be pushed further. Think of the ages which the molten earth required for its consolidation. But these vaster epochs lack sublimity through our inability to grasp them. They bewilder us, but they fail to make a solemn impression. The genesis of the mountains comes more within the scope of the intellect, and the majesty of the operation is enhanced by our partial ability to conceive it. In the falling of a rock from a mountain-head, in the shoot of an avalanche, in the plunge of a cataract, we often see more impressive illustrations of the power of gravity than in the motions of the stars. When the intellect has to intervene, and calculation is necessary to the building up of the conception, the expansion of the feelings ceases to be proportional to the magnitude of the phenomena.

———

I will here record a few other measurements executed on the Rosegg glacier: the line was staked out across the trunk formed by the junction of the Rosegg proper with the Tschierva glacier, a short distance below the rocky promontory called Agaliogs.

Rosegg Glacier.

No. of Stake.	Hourly Motion.
1	0·01 inch.
2	0·05 ,,
3	0·07 ,,
4	0·10 ,,
5	0·11 ,,
6	0·13 ,,
7	0·14 ,,
8	0·18 ,,
9	0·24 ,,
10	0·23 ,,
11	0·24 ,,

This is an extremely slowly moving glacier; the maximum motion hardly amounts to seven inches a day. Crevasses prevented us from continuing the line quite across the glacier.

X

RECENT EXPERIMENTS ON FOG-SIGNALS.[1]

THE care of its sailors is one of the first duties of a maritime people, and one of the sailor's greatest dangers is his proximity to the coast at night. Hence the idea of warning him of such proximity by beacon-fires placed sometimes on natural eminences and sometimes on towers built expressly for the purpose. Close to Dover Castle, for example, stands an ancient Pharos of this description.

As our marine increased greater skill was invoked, and lamps reinforced by parabolic reflectors poured their light upon the sea. Several of these lamps were sometimes grouped together so as to intensify the light, which at a little distance appeared as if it emanated from a single source. This 'catoptric' form of apparatus is still to some extent employed in our lighthouse-service, but for a long time past it has been more and more displaced by the great lenses devised by the illustrious Frenchman, Fresnel.

In a first-class 'dioptric' apparatus the light emanates from a lamp with several concentric wicks, the flame of which, being kindled by a very active draught, attains to great intensity. In fixed lights the lenses refract the rays issuing from the lamp so as to cause them to form a luminous sheet which grazes the

[1] A discourse delivered in the Royal Institution, March 22, 1878.

sea-horizon. In revolving lights the lenses gather up
the rays into distinct beams, resembling the spokes of a
wheel, which sweep over the sea and strike the eye of
the mariner in succession.

It is not for clear weather that the greatest strengthening of the light is intended, for here it is not needed.
Nor is it for densely foggy weather, for here it is ineffectual. But it is for the intermediate stages of hazy,
snowy, or rainy weather, in which a powerful light can
assert itself, while a feeble one is extinguished. The
usual first-order lamp is one of four wicks, but Mr.
Douglass, the able and indefatigable engineer of the
Trinity House, has recently raised the number of the
wicks to six, which produce a very noble flame. To Mr.
Wigham, of Dublin, we are indebted for the successful
application of gas to lighthouse illumination. In some
lighthouses his power varies from 28 jets to 108 jets,
while in the lighthouse of Galley Head three burners of
the largest size can be employed, the maximum number
of jets being 324. These larger powers are invoked only
in case of fog, the 28-jet burner being amply sufficient
for clear weather. The passage from the small burner
to the large, and from the large burner to the small, is
made with ease, rapidity, and certainty. This employment of gas is indigenous to Ireland, and the Board
of Trade has exercised a wise liberality in allowing
every facility to Mr. Wigham for the development of
his invention.

The last great agent employed in lighthouse illumination is electricity. It was in this Institution,
beginning in 1831, that Faraday proved the existence
and illustrated the laws of those induced currents which
in our day have received such astounding development.
In relation to this subject Faraday's words have a pro-
phetic ring. ' I have rather,' he writes in 1831, ' been

desirous of discovering new facts and new relations dependent on magneto-electric induction than of exalting the force of those already obtained, being assured that the latter would find their full development hereafter.' The labours of Holmes, of the Paris Alliance Company, of Wilde, and of Gramme, constitute a brilliant fulfilment of this prediction.

But, as regards the augmentation of power, the greatest step hitherto made was independently taken a few years ago by Dr. Werner Siemens and Sir Charles Wheatstone. Through the application of their discovery a machine endowed with an infinitesimal charge of magnetism may, by a process of accumulation at compound interest, be caused so to enrich itself magnetically as to cast by its performance all the older machines into the shade. The light now before you is that of a small machine placed downstairs, and worked there by a minute steam-engine. It is a light of about 1000 candles ; and for it, and for the steam-engine that works it, our members are indebted to the liberality of Dr. William Siemens, who in the most generous manner has presented the machine to this Institution. After an exhaustive trial at the South Foreland, machines on the principle of Siemens, but of far greater power than this one, have been recently chosen by the Elder Brethren of the Trinity House for the two light-houses at the Lizard Point.

Our most intense lights, including the six-wick lamp, the Wigham gas-light, and the electric light, being intended to aid the mariner in heavy weather, may be regarded, in a certain sense, as fog-signals. But fog, when thick, is intractable to light. The sun cannot penetrate it, much less any terrestrial source of illumination. Hence the necessity of employing sound-signals in dense fogs. Bells, gongs, horns, whistles, guns, and syrens have been used for this purpose ; but it is mainly, if not wholly, with explosive signals that we have now to deal.　The gun has been employed with useful effect at the North Stack, near Holyhead, on the Kish Bank near Dublin, at Lundy Island, and at other points on our coasts.　During the long, laborious, and I venture to think memorable series of observations conducted under the auspices of the Elder Brethren of the Trinity House at the South Foreland in 1872 and 1873, it was proved that a short $5\frac{1}{2}$-inch howitzer, firing 3 lbs. of powder, yielded a louder report than a long 18-pounder firing the same charge.　Here was a hint to be acted on by the Elder Brethren.　The effectiveness of the sound depended on the shape of the gun, and as it could not be assumed that in the howitzer we had hit accidentally upon the best possible shape, arrangements were made with the War Office for the construction of a gun specially calculated to produce the loudest sound attainable from the combustion of 3 lbs. of powder.　To prevent the unnecessary landward waste of the sound, the gun was furnished with a parabolic muzzle, intended to project the sound over the sea, where it was most needed. The construction of this gun was based on a searching series of experiments executed at Woolwich with small models, provided with muzzles of various kinds.　A drawing of the gun is annexed (p. 257).　It was constructed on the principle of the revolver, its various chambers being loaded and brought in rapid succession into the firing position.　The performance of the gun proved the correctness of the principles on which its construction was based.

An incidental point of some interest was decided by the earliest Woolwich experiments.　It had been a widely spread opinion among artillerists, that a bronze gun produces a specially loud report.　I doubted from the outset whether this would help us; and in a letter dated 22nd April, 1874, I ventured to express myself thus :—' The report of a gun, as affecting an observer close at hand, is made up of two factors—the sound due to the shock of the air by the violently expanding gas, and the sound derived from the vibrations of the gun, which, to some extent, rings like a bell.   This latter, I apprehend, will disappear at considerable distances.'

FIG. 6.

Breech-loading Fog-signal Gun, with Bell Mouth,[1] proposed by Major Maitland, R.A., Assistant Superintendent.

The result of subsequent trial, as reported by General Campbell, is, ' that the sonorous qualities of bronze are greatly superior to those of cast iron at short distances, but that the advantage lies with the baser metal at long ranges.' [2]

[1] The carriage of this gun has been modified in construction since this drawing was made.

[2] General Campbell assigns a true cause for this difference. The ring of the bronze gun represents so much energy withdrawn

Coincident with these trials of guns at Woolwich, gun-cotton was thought of as a probably effective sound-producer. From the first, indeed, theoretic considerations caused me to fix my attention persistently on this substance ; for the remarkable experiments of Mr. Abel, whereby its rapidity of combustion and violently explosive energy are demonstrated, seemed to single it out as a substance eminently calculated to fulfil the conditions necessary to the production of an intense wave of sound. What those conditions are we shall now more particularly enquire, calling to our aid a brief but very remarkable paper, published by Professor Stokes in the ' Philosophical Magazine ' for 1868.

The explosive force of gunpowder is known to depend on the sudden conversion of a solid body into an intensely heated gas. Now the work which the artillerist requires the expanding gas to perform is the displacement of the projectile, besides which it has to displace the air in front of the projectile, which is backed by the whole pressure of the atmosphere. Such, however, is not the work that we want our gunpowder to perform. We wish to transmute its energy not into the mere mechanical translation of either shot or air, but into vibratory motion. We want *pulses* to be formed which shall propagate themselves to vast distances through the atmosphere, and this requires a certain choice and management of the explosive material.

A sound-wave consists essentially of two parts—a condensation and a rarefaction. Now air is a very mobile fluid, and if the shock imparted to it lack due promptness, the wave is not produced. Consider the case of a common clock pendulum, which oscillates to

from the explosive force of the gunpowder. Further experiments would, however, be needed to place the superiority of the cast-iron gun at a distance beyond question.

and fro, and which might be expected to generate corresponding pulses in the air. When, for example, the bob moves to the right, the air to the right of it might be supposed to be condensed, while a partial vacuum might be supposed to follow the bob. As a matter of fact, we have nothing of the kind. The air particles in front of the bob retreat so rapidly, and those behind it close so rapidly in, that no sound-pulse is formed. The mobility of hydrogen, moreover, being far greater than that of air, a prompter action is essential to the formation of sonorous waves in hydrogen than in air. It is to this rapid power of readjustment, this refusal, so to speak, to allow its atoms to be crowded together or to be drawn apart, that Professor Stokes, with admirable penetration, refers the damping power, first described by Sir John Leslie, of hydrogen upon sound.

A tuning-fork which executes 256 complete vibrations in a second, if struck gently on a pad and held in free air, emits a scarcely audible note. It behaves to some extent like the pendulum bob just referred to. This feebleness is due to the prompt ' reciprocating flow ' of the air between the incipient condensations and rarefactions, whereby the formation of sound-pulses is forestalled. Stokes, however, has taught us that this flow may be intercepted by placing the edge of a card in close proximity to one of the corners of the fork. An immediate augmentation of the sound of the fork is the consequence.

The more rapid the shock imparted to the air, the greater is the fractional part of the energy of the shock converted into wave motion. And as different kinds of gunpowder vary considerably in their rapidity of combustion, it may be expected that they will also vary as producers of sound. This theoretic inference is completely verified by experiment. In a series of preliminary trials conducted at Woolwich on the 4th of June, 1875, the sound-producing powers of four different kinds of powder were determined. In the order of the size of their grains they bear the names respectively of Fine-grain (F. G.), Large-grain (L. G.), Rifle Large-grain (R. L. G.), and Pebble-powder (P.) (See annexed figures.) The charge in each case amounted to $4\frac{1}{2}$ lbs.;

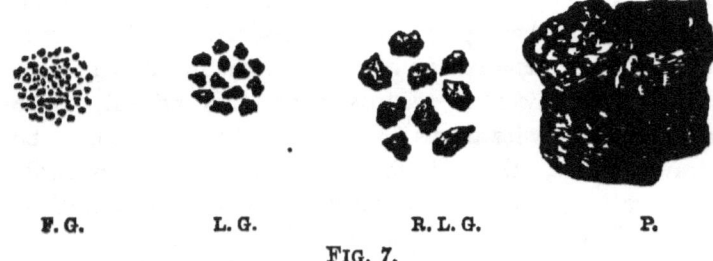

F. G.   L. G.   R. L. G.   P.

FIG. 7.

four 24-lb. howitzers being employed to fire the respective charges. There were eleven observers, all of whom, without a single dissentient, pronounced the sound of the fine-grain powder loudest of all. In the opinion of seven of the eleven the large-grain powder came next; seven also of the eleven placed the rifle large-grain third on the list; while they were again unanimous in pronouncing the pebble-powder the worst sound-producer These differences are entirely due to differences in the rapidity of combustion. All who have witnessed the performance of the 80-ton gun must have been surprised at the mildness of its thunder. To avoid the strain resulting from quick combustion, the powder employed is composed of lumps far larger than those of the pebble-powder above referred to. In the long tube of the gun these lumps of solid matter gradually resolve themselves into gas, which on issuing from the muzzle imparts a kind of push to the air, instead of the sharp shock necessary to form the condensation of an intensely sonorous wave.

These are some of the physical reasons why gun-cotton might be regarded as a promising fog-signal. Firing it as we have been taught to do by Mr. Abel, its explosion is more rapid than that of gunpowder. In its case the air particles, alert as they are, will not, it might be presumed, be able to slip from condensation to rarefaction with a rapidity sufficient to forestall the formation of the wave. On à *priori* grounds then, we are entitled to infer the effectiveness of gun-cotton, while in a great number of comparative experiments, stretching from 1874 to the present time, this inference has been verified in the most conclusive manner.

As regards explosive material, and zealous and accomplished help in the use of it, the resources of Woolwich Arsenal have been freely placed at the disposal of the Elder Brethren. General Campbell, General Younghusband, Colonel Fraser, Colonel Maitland, and other officers, have taken an active personal part in the investigation, and in most cases have incurred the labour of reducing and reporting on the observations. Guns of various forms and sizes have been invoked for gunpowder, while gun-cotton has been fired in free air and in the foci of parabolic reflectors.

On the 22nd of February, 1875, a number of small guns, cast specially for the purpose—some with plain, some with conical, and some with parabolic muzzles— firing 4 oz. of fine grain powder, were pitted against 4 oz. of gun-cotton detonated both in the open, and in the focus of a parabolic reflector.[1] The sound produced by the gun-cotton, reinforced by the reflector, was unanimously pronounced udest of all. With equal

[1] For charges of this weignt the reflector is of moderate size, and may be employed without fear of fracture.

unanimity, the gun-cotton detonated in free air was placed second in intensity. Though the same charge was used throughout, the guns differed notably among themselves, but none of them came up to the gun-cotton, either with or without the reflector. A second series, observed from a different distance on the same day, confirmed to the letter the foregoing result.

As a practical point, however, the comparative cost of gun-cotton and gunpowder has to be taken into account, though considerations of cost ought not to be stretched too far in cases involving the safety of human life. In the earlier experiments, where quantities of equal price were pitted against each other, the results were somewhat fluctuating. Indeed, the perfect manipulation of the gun-cotton required some preliminary discipline—promptness, certainty, and effectiveness of firing, augmenting as experience increased. As 1 lb. of gun-cotton costs as much as 3 lbs. of gunpowder, these quantities were compared together on the 22nd of February. The guns employed to discharge the gunpowder were a 12-lb. brass howitzer, a 24-lb. cast-iron howitzer, and the long 18-pounder employed at the South Foreland. The result was, that the 24-lb. howitzer, firing 3 lbs. of gunpowder, had a slight advantage over 1 lb. of gun-cotton detonated in the open; while the 12-lb. howitzer and the 18-pounder were both beaten by the gun-cotton. On the 2nd of May, on the other hand, the gun-cotton is reported as having been beaten by all the guns.

Meanwhile, the parabolic-muzzle gun, expressly intended for fog-signalling, was pushed rapidly forward, and on March 22 and 23, 1876, its power was tested at Shoeburyness. Pitted against it were a 16-pounder, a $5\frac{1}{2}$-inch howitzer, $1\frac{1}{2}$ lb. of gun-cotton detonated in the focus of a reflector (see annexed figure), and $1\frac{1}{2}$ lb. of

gun-cotton detonated in free air. On this occasion
nineteen different series of experiments were made, when
the new experimental gun, firing a 3-lb. charge, demonstrated its superiority over all guns previously employed
to fire the same charge. As regards the comparative

FIG. 8.

Gun-cotton Slab (1½ lb.) Detonated in the Focus of a Cast-iron
Reflector.

merits of the gun-cotton fired in the open, and the gunpowder fired from the new gun, the mean values of their
sounds were the same. Fired in the focus of the reflector, the gun-cotton clearly dominated over all the
other sound-producers.[1]

The whole of the observations here referred to were
embraced by an angle of about 70°, of which 50° lay on

[1] The reflector was fractured by the explosion, but it did good
service afterwards.

the one side and 20° on the other side of the line of fire.
The shots were heard by eleven observers on board the
'Galatea,' which took up positions varying from 2 miles
to 13¼ miles from the firing-point.   In all these observations, the reinforcing action of the reflector, and of
the parabolic muzzle of the gun, came into play.   But
the reinforcement of the sound in one direction implies
its withdrawal from some other direction, and accordingly it was found that at a distance of 5¼ miles from
the firing-point, and on a line including nearly an angle
of 90° with the line of fire, the gun-cotton in the open
beat the new gun; while behind the station, at distances
of 8½ miles and 13½ miles respectively, the gun-cotton
in the open beat both the gun and the gun-cotton in the
reflector.   This result is rendered more important by
the fact that the sound reached the Mucking Light, a
distance of 13¼ miles, against a light wind which was
blowing at the time.

Most, if not all, of our ordinary sound-producers
send forth waves which are not of uniform intensity
throughout.   A trumpet is loudest in the direction of
its axis.   The same is true of a gun.   A bell, with its
mouth pointed upwards or downwards, sends forth waves
which are far denser in the horizontal plane passing
through the bell than at an angular distance of 90° from
that plane.   The oldest bellhangers must have been
aware of the fact that the sides of the bell, and not its
mouth, emitted the strongest sound, their practice being
probably determined by this knowledge.   Our slabs of
gun-cotton also emit waves of different densities in different parts. It has occurred in the experiments at Shoeburyness that when the broad side of a slab was turned towards
the suspending wire of a second slab six feet distant, the
wire was cut by the explosion, while when the edge of
the slab was turned to the wire this never occurred.

To the circumstance that the broadsides of the slabs faced the sea is probably to be ascribed the remarkable fact observed on March 23, that in two directions, not far removed from the line of fire, the gun-cotton detonated in the open had a slight advantage over the new gun.

Theoretic considerations rendered .it probable that the shape and size of the exploding mass would affect the constitution of the wave of sound.   I did not think large rectangular slabs the most favourable shape, and accordingly proposed cutting a large slab into fragments of different sizes, and pitting them against each other. The differences between the sounds were by no means so great as the differences in the quantities of explosive material might lead one to expect.   The mean values of eighteen series of observations made on board the 'Galatea,' at distances varying from $1\frac{1}{4}$ mile to 4·8 miles, were as follows:—

Weights	4 oz.	6 oz.	9 oz.	12 oz.
Value of sound	3·12	3·34	4·0	4·03

These charges were cut from a slab of dry gun-cotton about $1\frac{3}{4}$ inch thick: they were squares and rectangles of the following dimensions:—4 oz., 2 inches by 2 inches; 6 oz., 2 inches by 3 inches; 9 oz., 3 inches by 3 inches; 12 oz., 2 inches by 6 inches.

The· numbers under the respective weights express the recorded value of the sounds.   They must be simply taken as a ready means of expressing the approximate relative intensity of the sounds as estimated by the ear. When we find a 9-oz. charge marked 4, and a 12-oz. charge marked 4·03, the two sounds may be regarded as practically equal in intensity, thus proving that an addition of 30 per cent. in the larger charges produces no sensible difference in the sound.   Were the sounds estimated by some physical means, instead of by the ear, the values of the sounds at the distances recorded would not, in my opinion, show a greater advance with the increase of material than that indicated by the foregoing numbers. Subsequent experiments rendered still more certain the effectiveness, as well as the economy, of the smaller charges of gun-cotton.

It is an obvious corollary from the foregoing experiments that on our 'nesses' and promontories, where the land is clasped on both sides for a considerable distance by the sea—where, therefore, the sound has to propagate itself rearward as well as forward—the use of the parabolic gun, or of the parabolic reflector, might be a disadvantage rather than an advantage. Here gun-cotton, exploded in the open, forms the most appropriate source of sound. This remark is especially applicable to such lightships as are intended to spread the sound all round them as from central foci. As a signal in rock lighthouses, where neither syren, steam-whistle, nor gun could be mounted; and as a handy fleet-signal, dispensing with the lumber of special signal-guns, the gun-cotton will prove invaluable. But in most of these cases we have the drawback that local damage may be done by the explosion. The lantern of the rock lighthouse might suffer from concussion near at hand, and though mechanical arrangements might be devised, both in the case of the lighthouse and of the ship's deck, to place the firing-point of the gun-cotton at a safe distance, no such arrangement could compete, as regards simplicity and effectiveness, with the expedient of *a gun-cotton rocket.* Had such a means of signalling existed at the Bishop's Rock lighthouse, the ill-fated 'Schiller' might have been warned of her approach to danger ten, or it may be twenty, miles before she reached the rock which wrecked her. Had the fleet possessed such a signal, instead of the ubiquitous but ineffectual whistle, the 'Iron Duke' and 'Vanguard' need never have come into collision.

It was the necessity of providing a suitable signal for rock lighthouses, and of clearing obstacles which cast an acoustic shadow, that suggested the idea of the gun-cotton rocket to Sir Richard Collinson, Deputy Master of the Trinity House. His idea was to place a disk or short cylinder of gun-cotton in the head of a rocket, the ascensional force of which should be employed to carry the disk to an elevation of 1000 feet or thereabouts, where by the ignition of a fuse associated with a detonator, the gun-cotton should be fired, sending its sound in all directions vertically and obliquely down upon earth and sea. The first attempt to realise this idea was made on July 18, 1876, at the firework manufactory of the Messrs. Brock, at Nunhead. Eight rockets were then fired, four being charged with 5 oz. and four with $7\frac{1}{2}$ oz. of gun-cotton. They ascended to a great height, and exploded with a very loud report in the air. On July 27, the rockets were tried at Shoeburyness. The most noteworthy result on this occasion was the hearing of the sounds at the Mouse Lighthouse, $8\frac{1}{2}$ miles E. by S., and at the Chapman Lighthouse, $8\frac{1}{2}$ miles W. by N.; that is to say, at opposite sides of the firing-point. It is worthy of remark that, in the case of the Chapman Lighthouse, land and trees intervened between the firing-point and the place of observation 'This,' as General Younghusband justly remarked at the time, 'may prove to be a valuable consideration if it should be found necessary to place a signal station in a position whence the sea could not be freely observed.' Indeed, the clearing of such obstacles was one of the objects which the inventor of the rocket had in view.

With reference to the action of the wind, it was thought desirable to compare the range of explosions produced near the surface of the earth with others produced at the elevation attainable by the gun-cotton rockets.　Wind and weather, however, are not at our command; and hence one of the objects of a series of experiments conducted on December 13, 1876, was not fulfilled.　It is worthy, however, of note that on this day, with smooth water and a calm atmosphere, the rockets were distinctly heard at a distance of 11·2 miles from the firing-point.　The quantity of gun-cotton employed was $7\frac{1}{2}$ oz.　On Thursday, March 8, 1877, these comparative experiments of firing at high and low elevations were pushed still further.　The gun-cotton near the ground consisted of $\frac{1}{2}$-lb. disks, suspended from a horizontal iron bar about $4\frac{1}{2}$ feet above the ground. The rockets carried the same quantity of gun-cotton in their heads, and the height to which they attained, as determined by a theodolite, was from 800 to 900 feet. The day was cold, with occasional squalls of snow and hail, the direction of the sound being at right angles to that of the wind.　Five series of observations were made on board the ' Vestal,' at distances varying from 3 to 6 miles.　The mean value of the explosions in the air exceeded that of the explosions near the ground by a small but sensible quantity.　At Windmill Hill, Gravesend, however, which was nearly to leeward, and $5\frac{1}{2}$ miles from the firing-point, in nineteen cases out of twenty-four the disk fired near the ground was loudest; while in the remaining five the rocket had the advantage.

Towards the close of the day the atmosphere became very serene.　A few distant cumuli sailed near the horizon, but the zenith and a vast angular space all round it were absolutely free from cloud　From the deck of the ' Galatea' a rocket was discharged, which reached a great elevation, and exploded with a loud report. Following this solid nucleus of sound was a continuous train of echoes, which retreated to a continually greater distance, dying gradually off into silence after seven seconds' duration. These echoes were of the same character as those so frequently noticed at the South Foreland in 1872-73, and called by me 'aërial echoes.'

On the 23rd of March the experiments were resumed, the most noteworthy results of that day's observations being that the sounds were heard at Tillingham, 10 miles to the N.E.; at West Mersea, $15\frac{3}{4}$ miles to the N.E. by E.; at Brightlingsea, $17\frac{1}{4}$ miles to the N.E.; and at Clacton Wash, $20\frac{1}{2}$ miles to the N.E. by $\frac{1}{2}$ E. The wind was blowing at the time from the S.E. Some of these sounds were produced by rockets, some by a 24-lb. howitzer, and some by an 8-inch Maroon.

In December, 1876, Mr. Gardiner, the managing director of the Cotton-powder Company, had proposed a trial of this material against the gun-cotton. The density of the cotton he urged was only 1·03, while that of the powder was 1·70. A greater quantity of explosive material being thus compressed into the same volume, Mr. Gardiner thought that a greater sonorous effect must be produced by the powder. At the instance of Mr. Mackie. who had previously gone very thoroughly into the subject, a Committee of the Elder Brethren visited the cotton-powder manufactory, on the banks of the Swale, near Faversham, on the 16th of June, 1877. The weights of cotton-powder employed were 2 oz., 8 oz., 1 lb., and 2 lbs., in the form of rockets and of signals fired a few feet above the ground. The experiments throughout were arranged and conducted by Mr. Mackie. Our desire on this occasion was to get as near to windward as possible, but the Swale and other obstacles limited our distance to 1½ mile. We stood here E.S.E. from the firing-point while the wind blew fresh from the N.E.

The cotton-powder yielded a very effective report. The rockets in general had a slight advantage over the same quantities of material fired near the ground. The loudness of the sound was by no means proportional to the quantity of the material exploded, 8 oz. yielding very nearly as loud a report as 1 lb. The 'aërial echoes,' which invariably followed the explosion of the rockets, were loud and long-continued.

On the 17th of October, 1877, another series of experiments with howitzers and rockets was carried out at Shoeburyness. The charge of the howitzer was 3 lbs. of L. G. powder. The charges of the rockets were 12 oz., 8 oz., 4 oz., and 2 oz. of gun-cotton respectively. The gun and the four rockets constituted a series, and eight series were fired during the afternoon of the 17th. The observations were made from the 'Vestal' and the 'Galatea,' positions being successively assumed which permitted the sound to reach the observers with the wind, against the wind, and across the wind. The distance of the 'Galatea' varied from 3 to 7 miles, that of the 'Vestal,' which was more restricted in her movements, being 2 to 3 miles. Briefly summed up, the result is that the howitzer, firing a 3-lb. charge, which it will be remembered was our best gun at the South Foreland, was beaten by the 12-oz. rocket, by the 8-oz. rocket, and by the 4-oz. rocket. The 2-oz. rocket alone fell behind the howitzer.

It is worth while recording the distances at which some of the sounds were heard on the day now referred to:—

1. Leigh	.	.	6½ miles W.N.W.	24 out of 40 sounds heard.
2. Girdler Light-				
vessel .	.	12	„ S.E. by E. 5	„
3. Reculvers	.	17½	„ S.E. by S. 18	„
4. St. Nicholas	.	20	„ S.E. . 3	„
5. Epple Bay	.	22	„ S.E. by E. 19	„
6. Westgate	.	23	„ S.E. by E. 9	„
7. Kingsgate		25	„ S.E. by E. 8	„

The day was cloudy, with occasional showers of drizzling rain; the wind about N.W. by N. all day; at times squally, rising to a force of 6 or 7 and sometimes dropping to a force of 2 or 3. The station at Leigh excepted, all these places were to leeward of Shoebury-ness. At four other stations to leeward, varying in distance from 15½ to 24½ miles, nothing was heard, while at eleven stations to windward, varying from 8 to 26 miles, the sounds were also inaudible. It was found, indeed, that the sounds proceeding directly against the wind did not penetrate much beyond 3 miles.

On the following day, viz. the 18th October, we proceeded to Dungeness with the view of making a series of strict comparative experiments with gun-cotton and cotton-powder. Rockets containing 8 oz., 4 oz., and 2 oz. of gun-cotton had been prepared at the Royal Arsenal; while others, containing similar quantities of cotton-powder, had been supplied by the Cotton-powder Company at Faversham. With these were compared the ordinary 18-pounder gun, which happened to be mounted at Dungeness, firing the usual charge of 3 lbs. of powder, and a syren.

From these experiments it appeared that the gun-cotton and cotton-powder were practically equal as producers of sound.

The effectiveness of small charges was illustrated in a very striking manner, only a single unit separating the numerical value of the 8-oz. rocket from that of the

2-oz. rocket. The former was recorded as 6·9 and the
latter as 5·9, the value of the 4-oz. rocket being intermediate between them. These results were recorded by
a number of very practised observers on board the
'Galatea.' They were completely borne out by the
observations of the Coastguard, who marked the value
of the 8-oz rocket 6·1, and that of the 2-oz. rocket 5·2.
The 18-pounder gun fell far behind all the rockets, a
result, possibly, to be in part ascribed to the imperfec-
tion of the powder. The performance of the syren was,
on the whole, less satisfactory than that of the rocket.
The instrument was worked, not by steam of 70 lbs.
pressure, as at the South Foreland, but by compressed
air, beginning with 40 lbs. and ending with 30 lbs.
pressure. The trumpet was pointed to windward, and
in the axis of the instrument the sound was about as
effective as that of the 8-oz. rocket. But in a direction
at right angles to the axis, and still more in the rear of
this direction, the syren fell very sensibly behind even
the 2-oz. rocket.

These are the principal comparative trials made between the gun-cotton rocket and other fog-signals ; but
they are not the only ones. On the 2nd of August,
1877, for example, experiments were made at Lundy
Island with the following results. At 2 miles distant
from the firing-point, with land intervening, the 18-
pounder, firing a 3-lb. charge, was quite unheard.
Both the 4-oz. rocket and the 8-oz. rocket, however,
reached an elevation which commanded the acoustic
shadow, and yielded loud reports. When both were in
view the rockets were still superior to the gun. On
the 6th of August, at St. Ann's, the 4-oz. and 8-oz.
rockets proved superior to the syren. On the Shambles
Light-vessel, when a pressure of 13 lbs. was employed
to sound the syren, the rockets proved greatly superior

to that instrument.   Proceeding along the sea margin
at Flamboro' Head, Mr. Edwards states that at a distance of 1¼ mile, with the 18-pounder previously used
as a fog-signal hidden behind the cliffs, its report was
quite unheard, while the 4-oz. rocket, rising to an elevation which brought it clearly into view, yielded a powerful sound in the face of an opposing wind.

On the evening of February 9th, 1877, a remarkable
series of experiments were made by Mr. Prentice at
Stowmarket with the gun-cotton rocket.   From the
report with which he has kindly furnished me I extract
the following particulars.   The first column in the
annexed statement contains the name of the place of
observation, the second its distance from the firing-point,
and the third the result observed :—

Stoke Hill, Ipswich	. 10 miles	Rockets clearly seen and sounds distinctly heard 53 seconds after the flash.
Melton . . .	15 „	Signals distinctly heard. Thought at first that sounds were reverberated from the sea.
Framlingham . .	18 „	Signals very distinctly heard, both in the open air and in a closed room. Wind in favour of sound.
Stratford. St. Andrews .	19 „	Reports loud ; startled pheasants in a cover close by.
Tuddenham. St. Martin	10 „	Reports very loud ; rolled away like thunder.
Christ Church Park.	. 11 „	Report arrived a little more than a minute after flash.
Nettlestead Hall .	. 6 „	Distinct in every part of ob-server's house. Very loud in the open air.
Bildestone . .	. 6 · „	Explosion very loud, wind against sound.
Nacton . . .	. 14 „	Reports quite distinct — mis-taken by inhabitants for claps of thunder.

Aldboro' .	.	.	. 25 miles	Rockets seen through a very hazy atmosphere; a rumbling detonation heard.
Capel Mills .	.	.	. 11 „	Reports heard within and without the observer's house. Wind opposed to sound.
Lawford .	.	.	. 15½ „	Reports distinct: attributed to distant thunder.

In the great majority of these cases, the direction of the sound enclosed a large angle with the direction of the wind. In some cases, indeed, the two directions were at right angles to each other. It is needless to dwell for a moment on the advantage of possessing a signal commanding ranges such as these.

The explosion of substances in the air, after having been carried to a considerable elevation by rockets, is a familiar performance. In 1873, moreover, the Board of Trade proposed a light-and-sound rocket as a signal of distress, which proposal was subsequently realized, but in a form too elaborate and expensive for practical use. The idea of a gun-cotton rocket fit for signalling in fogs is, I believe, wholly due to Sir Richard Collinson, the Deputy Master of the Trinity House. Thanks to the skilful aid given by the authorities of Woolwich, by Mr. Prentice, and Mr. Brock, that idea is now an accomplished fact; a signal of great power, handiness, and economy, being thus placed at the service of our mariners. Not only may the rocket be applied in association with lighthouses and lightships, but in the Navy also it may be turned to important account. Soon after the loss of the 'Vanguard' I ventured to urge upon an eminent naval officer the desirability of having an organized code of fog-signals for the fleet. He shook his head doubtingly, and referred to the difficulty of finding room for signal guns. The gun-cotton rocket completely surmounts this difficulty. It is manipulated with ease and rapidity, while its discharges may be so grouped and combined as to give a most important extension to the voice of the admiral in command. It is needless to add that at any point upon our coasts, or upon any other coast, where its establishment might be desirable, a fog-signal station might be extemporised without difficulty.

I have referred more than once to the train of echoes which accompanied the explosion of gun-cotton in free air, speaking of them as similar in all respects to those which were described for the first time in my Report on Fog-signals, addressed to the Corporation of Trinity House in 1874.[1] To these echoes I attached a fundamental significance. There was no visible reflecting surface from which they could come. On some days, with hardly a cloud in the air and hardly a ripple on the sea, they reached a magical intensity. As far as the sense of hearing could judge, they came from the body of the air in front of the great trumpet which produced them. The trumpet blasts were five seconds in duration, but long before the blast had ceased the echoes struck in, adding their strength to the primitive note of the trumpet. After the blast had ended the echoes continued, retreating further and further from the point of observation, and finally dying away at great distances. The echoes were perfectly continuous as long as the sea was clear of ships, 'tapering' by imperceptible gradations into absolute silence. But when a ship happened to throw itself athwart the course of the sound, the echo from the broadside of the vessel was returned as a shock which rudely interrupted the continuity of the dying atmospheric music.

[1] See also 'Philosophical Transactions' for 1874, p. 183.

These echoes have been ascribed to reflection from the crests of the sea-waves. But this hypothesis is negatived by the fact, that the echoes were produced in great intensity and duration when no waves existed —when the sea, in fact, was of glassy smoothness. It has been also shown that the direction of the echoes depended not on that of waves, real or assumed, but on the direction of the axis of the trumpet. Causing that axis to traverse an arc of 210°, and the trumpet to sound at various points of the arc, the echoes were always, at all events in calm weather, returned from that portion of the atmosphere towards which the trumpet was directed. They could not, under the circumstances, come from the glassy sea; while both their variation of direction and their perfectly continuous fall into silence, are irreconcilable with the notion that they came from fixed objects on the land. They came from that portion of the atmosphere into which the trumpet poured its maximum sound, and fell in intensity as the direct sound penetrated to greater atmospheric distances.

The day on which our latest observations were made was particularly fine. Before reaching Dungeness, the smoothness of the sea and the serenity of the air caused me to test the echoing power of the atmosphere. A single ship lay about half a mile distant between us and the land. The result of the proposed experiment was clearly foreseen. It was this. The rocket being sent up, it exploded at a great height; the echoes retreated in their usual fashion, becoming less and less intense as the distances of the invisible surfaces of reflection from the observers increased. About five seconds after the explosion, a single loud shock was sent back to us from the side of the vessel lying between us and the land. Obliterated for a moment by this more intense echo, the aërial reverberation continued its retreat, dying away into silence in two or three seconds afterwards.[1]

I have referred to the firing of an 8-oz. rocket from the deck of the 'Galatea' on March 8, 1877, stating the duration of its echoes to be seven seconds. Mr. Prentice, who was present at the time, assured me that in his experiments similar echoes had been frequently heard of more than twice this duration. The ranges of his sounds alone would render this result in the highest degree probable.

To attempt to interpret an experiment which I have not had an opportunity of repeating, is an operation of some risk; and it is not without a consciousness of this that I refer here to a result announced by Professor Joseph Henry, which he considers adverse to the notion of aërial echoes. He took the trouble to point the trumpet of a syren towards the zenith, and found that when the syren was sounded no echo was returned. Now the reflecting surfaces which give rise to these echoes are for the most part due to differences of temperature between sea and air. If, through any cause, the air above be chilled, we have descending streams—if the air below be warmed, we have ascending streams as the initial cause of atmospheric flocculence. A sound proceeding vertically does not cross the streams, nor impinge upon the reflecting surfaces, as does a sound proceeding horizontally across them. Aërial echoes, therefore, will not accompany the vertical sound as they accompany the horizontal one. The experiment, as I interpret it, is not opposed to the theory of these echoes which I have ventured to enunciate. But, as I have indicated, not only to see but to vary such an

---

[1] The echoes of the gun fired on shore this day were very brief; those of the 12-oz. gun-cotton rocket were 12″ and those of the 8-oz. cotton-powder rocket 11″ in duration.

experiment is a necessary prelude to grasping its full significance.

In a paper published in the 'Philosophical Transactions' for 1876, Professor Osborne Reynolds refers to these echoes in the following terms:—' Without attempting to explain the reverberations and echoes which have been observed, I will merely call attention to the fact that in no case have I heard any attending the reports of the rockets,[1] although they seem to have been invariable with the guns and pistols. These facts suggest that the echoes are in some way connected with the direction given to the sound. They are caused by the voice, trumpets, and the syren, all of which give direction to the sound; but I am not aware that they have ever been observed in the case of a sound which has no direction of greatest intensity.' The reference to the voice, and other references in his paper, cause me to think that, in speaking of echoes, Professor Osborne Reynolds and myself are dealing with different phenomena. Be that as it may, the foregoing observations render it perfectly certain that the condition as to direction here laid down is not necessary to the production of the echoes.

There is not a feature connected with the aërial echoes which cannot be brought out by experiments in the air of the laboratory. I have recently made the following experiment:—A rectangle, x y (p. 279), 22 inches by 12, was crossed by twenty-three brass tubes (half the number would suffice and only eleven are shown in the figure), each having a slit along it from which gas can issue. In this way twenty-three low flat flames were obtained. A sounding reed *a* fixed in a

---

[1] These carried 12 oz. of gunpowder, which has been found by Col. Fraser to require an iron case to produce an effective explosion.

short tube was placed at one end of the rectangle, and
a 'sensitive flame,'[1] f, at some distance beyond the
other end.   When the reed sounded, the flame in front
of it was violently agitated, and roared boisterously.
Turning on the gas, and lighting it as it issued from
the slits, the air above the flames became so heterogeneous that the sensitive flame was instantly stilled,
rising from a height of 6 inches to a height of 18
inches.   Here we had the acoustic opacity of the air
in front of the South Foreland strikingly imitated.[2]
Turning off the gas, and removing the sensitive flame
to f', some distance behind the reed, it burned there
tranquilly, though the reed was sounding.   Again

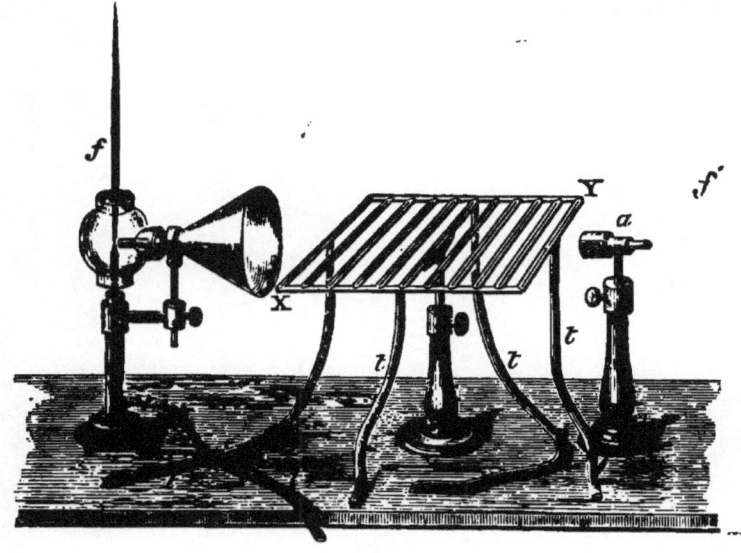

FIG. 9.

lighting the gas as it issued from the brass tubes,
the sound reflected from the heterogeneous air threw

[1] Fully described in my 'Lectures on Sound,' 3rd edition, p. 227.
[2] Lectures on Sound, 3rd ed., p. 268.

the sensitive flame into violent agitation. Here we had imitated the aërial echoes heard when standing behind the syren-trumpet at the South Foreland. The experiment is extremely simple, and in the highest degree impressive.

———————

The explosive rapidity of dynamite marks it as a substance specially suitable for the production of sound. At the suggestion of Professor Dewar, Mr. McRoberts has carried out a series of experiments on dynamite, with extremely promising results. Immediately after the delivery of the foregoing lecture I was informed that Mr. Brock proposed the employment of dynamite in the Collinson rocket.

XI.

ON THE STUDY OF PHYSICS.[1]

I HOLD in my hand an uncorrected proof of the sylla-
bus of this course of lectures, and the title of
the present lecture is there stated to be ' On the Importance of the Study of Physics as a *Means* of Education.'
The corrected proof, however, contains the title :—' On
the Importance of the Study, of Physics as a *Branch*
of Education.' Small as this editorial alteration may
seem, the two words suggest two radically distinct
modes of viewing the subject before us. The term
Education is sometimes applied to a single faculty or
organ, and if we know wherein the education of a
single faculty consists, this will help us to clearer
notions regarding the education of the sum of all the
faculties, or of the mind. When, for example, we speak
of the education of the voice, what do we mean?
There are certain membranes at the top of the
windpipe which throw into vibration the air forced
between them from the lungs, thus producing musical
sounds. These membranes are, to some extent, under
the control of the will, and it is found that they can be
so modified by exercise as to produce notes of a clearer
and more melodious character. This exercise we call
the education of the voice. We may choose for our

[1] From a lecture delivered in the Royal Institution of Great
Britain in the Spring of 1854.

exercise songs new or old, festive or solemn; the education of the voice being the object aimed at, the songs
may be regarded as the means by which this education
is accomplished. I think this expresses the state of
the case more clearly than if we were to call the songs
a branch of education. Regarding also the education
of the human mind as the improvement and development of the mental faculties, I shall consider the study
of Physics as a means towards the attainment of this
end. From this point of view, I degrade Physics into
an implement of culture, and this is my deliberate
design.

The term Physics, as made use of in the present
Lecture, refers to that portion of natural science which
lies midway between astronomy and chemistry. The
former, indeed, is Physics applied to 'masses of enormous weight,' while the latter is Physics applied to
atoms and molecules. The subjects of Physics proper
are therefore those which lie nearest to human perception:—light and heat, colour, sound, motion, the
loadstone, electrical attractions and repulsions, thunder
and lightning, rain, snow, dew, and so forth. Our
senses stand between these phenomena and the reasoning
mind. We observe the fact, but are not satisfied with
the mere act of observation: the fact must be accounted
for—fitted into its position in the line of cause and
effect. Taking our facts from Nature we transfer
them to the domain of thought: look at them, compare
them, observe their mutual relations and connexions,
and bringing them ever clearer before the mental eye,
finally alight upon the cause which unites them. This
is the last act of the mind, in this centripetal direction—
in its progress from the multiplicity of facts to the
central cause on which they depend. But, having
guessed the cause, we are not yet contented. We set

out from the centre and travel in the other direction.
If the guess be true, certain consequences must follow
from it, and we appeal to the law and testimony of
experiment whether the thing is so.   Thus is the
circuit of thought completed,—from without inward,
from multiplicity to unity, and from within outward,
from unity to multiplicity.   In thus traversing both
ways the line between cause and effect, all our reasoning powers are called into play.   The mental effort
involved in these processes may be compared to those
exercises of the body which invoke the co-operation of
every muscle, and thus confer upon the whole frame the
benefits of healthy action.

The first experiment a child makes is a physical
experiment: the suction-pump is but an imitation of
the first act of every new-born infant.   Nor do I think
it calculated to lessen that infant's reverence, or to
make him a worse citizen, when his riper experience
shows him that the atmosphere was his helper in extracting the first draught from his mother's breast.
The child grows, but is still an experimenter : he grasps
at the moon, and his failure teaches him to respect
distance.   At length his little fingers acquire sufficient
mechanical tact to lay hold of a spoon.   He thrusts
the instrument into his mouth, hurts his gums, and
thus learns the impenetrability of matter.   He lets the
spoon fall, and jumps with delight to hear it rattle
against the table.   The experiment made by accident
is repeated with intention, and thus the young student
receives his first lessons upon sound and gravitation.
There are pains and penalties, however, in the path of
the enquirer: he is sure to go wrong, and Nature is
just as sure to inform him of the fact.   He falls downstairs, burns his fingers, cuts his hand, scalds his tongue,
and in this way learns the conditions of his physical

well being.   This is Nature's way of proceeding, and it
is wonderful what progress her pupil makes.   His
enjoyments for a time are physical, and the con-
fectioner's shop occupies the foreground of human
happiness ; but the blossoms of a finer life are already
beginning to unfold themselves, and the relation of
cause and effect dawns upon the boy.   He begins to
see that the present condition of things is not final,
but depends upon one that has gone before, and will be
succeeded by another.   He becomes a puzzle to himself ;
and to satisfy his newly-awakened curiosity, asks all
manner of inconvenient questions.   The needs and
tendencies of human nature express themselves through
these early yearnings of the child.   As thought ripens,
he desires to know the character and causes of the
phenomena presented to his observation ; and unless
this desire has been granted for the express purpose of
having it repressed, unless the attractions of natural
phenomena be like the blush of the forbidden fruit,
conferred merely for the purpose of exercising our self-
denial in letting them alone ; we may fairly claim for
the study of Physics the recognition that it answers to an
impulse implanted by nature in the constitution of man.

A few days ago, a Master of Arts, who is still a
young man, and therefore the recipient of a modern
education, stated to me that until he had reached the
age of twenty years he had never been taught anything
whatever regarding natural phenomena, or natural law.
Twelve years of his life previously had been spent
exclusively among the ancients.   The case, I regret to
say, is typical.   Now, we cannot, without prejudice to
humanity, separate the present from the past.   The
nineteenth century strikes its roots into the centuries
gone by, and draws nutriment from them.   The world
cannot afford to lose the record of any great deed or

utterance; for such are prolific throughout all time. We cannot yield the companionship of our loftier brothers of antiquity,—of our Socrates and Cato,—whose lives provoke us to sympathetic greatness across the interval of two thousand years. As long as the ancient languages are the means of access to the ancient mind, they must ever be of priceless value to humanity; but surely these avenues might be kept open without making such sacrifices as that above referred to, universal. We have conquered and possessed ourselves of continents of land, concerning which antiquity knew nothing; and if new continents of thought reveal themselves to the exploring human spirit, shall we not possess them also? In these latter days, the study of Physics has given us glimpses of the methods of Nature which were quite hidden from the ancients, and we should be false to the trust committed to us, if we were to sacrifice the hopes and aspirations of the Present out of deference to the Past.

The bias of my own education probably manifests itself in a desire I always feel to seize upon every possible opportunity of checking my assumptions and conclusions by experience. In the present case, it is true, your own consciousness might be appealed to in proof of the tendency of the human mind to inquire into the phenomena presented to it by the senses; but I trust you will excuse me if, instead of doing this, I take advantage of the facts which have fallen in my way through life, referring to your judgment to decide whether such facts are truly representative and general, and not merely individual and local.

At an agricultural college in Hampshire, with which I was connected for some time, and which is now converted into a school for the general education of youth, a Society was formed among the boys, who met weekly for the purpose of reading reports and papers upon various subjects.   The Society had its president and treasurer; and abstracts of its proceedings were published in a little monthly periodical issuing from the school press.   One of the most remarkable features of these weekly meetings was, that after the general business had been concluded, each member enjoyed the right of asking questions on any subject on which he desired information.   The questions were either written out previously in a book, or, if a question happened to suggest itself during the meeting, it was written upon a slip of paper and handed in to the Secretary, who afterwards read all the questions aloud.   A number of teachers were usually present, and they and the boys made a common stock of their wisdom in furnishing replies.   As might be expected from an assemblage of eighty or ninety boys, varying from eighteen to eight years old, many odd questions were proposed.   To the mind which loves to detect in the tendencies of the young the instincts of humanity generally, such questions are not without a certain philosophic interest, and I have therefore thought it not derogatory to the present course of Lectures to copy a few of them, and to introduce them here.   They run as follows :—

What are the duties of the Astronomer Royal?

What is frost ?

Why are thunder and lightning more frequent in summer than in winter?

What occasions falling stars ?

What is the cause of the sensation called ' pins and needles ' ?

What is the cause of waterspouts ?

What is the cause of hiccup ?

If a towel be wetted with water, why does the wet portion become darker than before ?

What is meant by Lancashire witches?

Does the dew rise or fall?

What is the principle of the hydraulic press?

Is there more oxygen in the air in summer than in winter?

What are those rings which we see round the gas and sun?

What is thunder?

How is it that a black hat can be moved by forming round it a magnetic circle, while a white hat remains stationary?

What is the cause of perspiration?

Is it true that men were once monkeys?

What is the difference between the *soul* and the *mind*?

Is it contrary to the rules of Vegetarianism to eat eggs?

In looking over these questions, which were wholly unprompted, and have been copied almost at random from the book alluded to, we see that many of them are suggested directly by natural objects, and are not such as had an interest conferred on them by previous culture. Now the fact is beyond the boy's control, and so certainly is the desire to know its cause. The sole question then is, whether this desire is to be gratified or not. Who created the fact? Who implanted the desire? Certainly not man. Who then will undertake to place himself between the desire and its fulfilment, and proclaim a divorce between them? Take, for example, the case of the wetted towel, which at first sight appears to be one of the most unpromising questions in the list. Shall we tell the proposer to repress his curiosity, as the subject is improper for him to know, and thus interpose our wisdom to rescue the boy from the consequences of a wish which acts to his prejudice? Or,

recognising the propriety of the question, how shall we
answer it?   It is impossible to answer it without
reference to the laws of optics—without making the
boy to some extent a natural philosopher. You may
say that the effect is due to the reflection of light
at the common surface of two media of different
refractive indices.   But this answer presupposes on the
part of the boy a knowledge of what reflection and
refraction are, or reduces you to the necessity of
explaining them.

On looking more closely into the matter, we find
that our wet towel belongs to a class of phenomena
which have long excited the interest of philosophers.
The towel is white for the same reason that snow is
white, that foam is white, that pounded granite or
glass is white, and that the salt we use at table is
white.   On quitting one medium and entering another,
a portion of light is always reflected, but on this condition—the media must possess different refractive indices.
Thus, when we immerse a bit of glass in water, light
is reflected from the common surface of both, and it is
this light which enables us to see the glass.   But when
a transparent solid is immersed in a liquid of the same
refractive index as itself, it immediately disappears. I
remember once dropping the eyeball of an ox into
water; it vanished as if by magic, with the exception
of the crystalline lens, and the surprise was so great
as to cause a bystander to suppose that the vitreous
humour had been instantly dissolved. This, however,
was not the case, and a comparison of the refractive
index of the humour with that of water cleared up
the whole matter. The indices were identical, and
hence the light pursued its way through both as if they
formed one continuous mass.

In the case of snow, powdered quartz, or salt, we

have a transparent solid mixed with air. At every transition from solid to air, or from air to solid, a portion of light is reflected, and this takes place so often that the light is wholly intercepted. Thus from the mixture of two transparent bodies we obtain an opaque one. Now the case of the towel is precisely similar. The tissue is composed of semi-transparent vegetable fibres, with the interstices between them filled with air ; repeated reflection takes place at the limiting surfaces of air and fibre, and hence the towel becomes opaque like snow or salt. But if we fill the interstices with water, we diminish the reflection ; a portion of the light is transmitted, and the darkness of the towel is due to its increased transparency. Thus the deportment of various minerals, such as hydrophane and tabasheer, the transparency of tracing paper used by engineers, and many other considerations of the highest scientific interest, are involved in the simple enquiry of this unsuspecting little boy.

Again, take the question regarding the rising or falling of the dew—a question long agitated, and finally set at rest by the beautiful researches of Wells. I do not think that any boy of average intelligence will be satisfied with the simple answer that the dew falls. He will wish to learn how you know that it falls, and, if acquainted with the notions of the middle ages, he may refer to the opinion of Father Laurus, that a goose egg filled in the morning with dew and exposed to the sun, will rise like a balloon—a swan's egg being better for the experiment than a goose egg. It is impossible to give the boy a clear notion of the beautiful phenomenon to which his question refers, without first making him acquainted with the radiation and conduction of heat. Take, for example, a blade of grass, from which one of these orient pearls is depending,

During the day the grass, and the earth beneath it, possess a certain amount of warmth imparted by the sun; during a serene night, heat is radiated from the surface of the grass into space, and to supply the loss, there is a flow of heat from the earth to the blade. Thus the blade loses heat by radiation, and gains heat by conduction. Now, in the case before us, the power of radiation is great, whereas the power of conduction is small; the consequence is that the blade loses more than it gains, and hence becomes more and more refrigerated. The light vapour floating around the surface so cooled is condensed upon it, and there accumulates to form the little pearly globe which we call a dew-drop.

Thus the **boy** finds the simple and homely fact which addressed his senses to be the outcome and flower of the deepest laws. The fact becomes, in a measure, sanctified as an object of thought, and invested for him with a beauty for evermore. He thus learns that things which, at first sight, seem to stand isolated and without apparent brotherhood in Nature are organically united, and finds the detection of such analogies a source of perpetual delight. To enlist pleasure on the side of intellectual performance is a point of the utmost importance; for the exercise of the mind, like that of the body, depends for its value upon the spirit in which it is accomplished. Every physician knows that something more than mere mechanical motion is comprehended under the idea of healthful exercise—that, indeed, being most healthful which makes us forget all ulterior ends in the mere enjoyment of it. What, for example, could be substituted for the action of the playground, where the boy plays for the mere love of playing, and without reference to physiological laws; while kindly Nature accomplishes her ends uncon-

sciously, and makes his very indifference beneficial to
him. You may have more systematic motions, you
may devise means for the more perfect traction of each
particular muscle, but you cannot create the joy and
gladness of the game, and where these are absent, the
charm and the health of the exercise are gone. The
case is similar with the education of the mind.

The study of Physics, as already intimated, consists
of two processes, which are complementary to each
other—the tracing of facts to their causes, and the
logical advance from the cause to the fact. In the
former process, called *induction*, certain moral qualities
come into play. The first condition of success is patient
industry, an honest receptivity, and a willingness to
abandon all preconceived notions, however cherished, if
they be found to contradict the truth. Believe me, a
self-renunciation which has something lofty in it, and
of which the world never hears, is often enacted in the
private experience of the true votary of science. And
if a man be not capable of this self-renunciation—this
loyal surrender of himself to Nature and to fact, he
lacks, in my opinion, the first mark of a true philosopher. Thus the earnest prosecutor of science, who
does not work with the idea of producing a sensation
in the world, who loves the truth better than the
transitory blaze of to-day's fame, who comes to his task
with a single eye, finds in that task an indirect means
of the highest moral culture. And although the virtue
of the act depends upon its privacy, this sacrifice of
self, this upright determination to accept the truth, no
matter how it may present itself—even at the hands of
a scientific foe, if necessary—carries with it its own
reward. When prejudice is put under foot and the
stains of personal bias have been washed away—when a
man consents to lay aside his vanity and to become

Nature's organ—his elevation is the instant consequence of his humility. I should not wonder if my remarks provoked a smile, for they seem to indicate that I regard the man of science as a heroic, if not indeed an angelic, character; and cases may occur to you which indicate the reverse. You may point to the quarrels of scientific men, to their struggles for priority, to that unpleasant egotism which screams around its little property of discovery like a scared plover about its young. I will not deny all this; but let it be set down to its proper account, to the weakness—or, if you will—to the selfishness of Man, but not to the charge of Physical Science.

The second process in physical investigation is *deduction*, or the advance of the mind from fixed principles to the conclusions which flow from them. The rules of logic are the formal statement of this process, which, however, was practised by every healthy mind before ever such rules were written. In the study of Physics, induction and deduction are perpetually wedded to each other. The man observes, strips facts of their peculiarities of form, and tries to unite them by their essences; having effected this, he at once deduces, and thus checks his induction. Here the grand difference between the methods at present followed, and those of the ancients, becomes manifest. They were one-sided in these matters: they omitted the process of induction, and substituted conjecture for observation. They could never, therefore, fulfil the mission of Man to 'replenish the earth, and subdue it.' The subjugation of Nature is only to be accomplished by the penetration of her secrets and the patient mastery of her laws. This not only enables us to protect ourselves from the hostile action of natural forces, but makes them our slaves. By the study of Physics we have indeed opened to us treasuries of power of which antiquity never dreamed. But while we lord it over Matter, we have thereby become better acquainted with the laws of Mind; for to the mental philosopher the study of Physics furnishes a screen against which the human spirit projects its own image, and thus becomes capable of self-inspection.

Thus, then, as a means of intellectual culture, the study of Physics exercises and sharpens observation : it brings the most exhaustive logic into play: it compares, abstracts, and generalizes, and provides a mental scenery appropriate to these processes. The strictest precision of thought is everywhere enforced, and prudence, foresight, and sagacity are demanded. By its appeals to experiment, it continually checks itself, and thus walks on a foundation of facts. Hence the exercise it invokes does not end in a mere game of intellectual gymnastics, such as the ancients delighted in, but tends to the mastery of Nature. This gradual conquest of the external world, and the consciousness of augmented strength which accompanies it, render the study of Physics as delightful as it is important.

With regard to the effect on the imagination, certain it is that the cool results of physical induction furnish conceptions which transcend the most daring flights of that faculty. Take for example the idea of an all-pervading ether which transmits a tingle, so to speak, to the finger ends of the universe every time a street lamp is lighted. The invisible billows of this ether can be measured with the same ease and certainty as that with which an engineer measures a base and two angles, and from these finds the distance across the Thames. Now it is to be confessed that there may be just as little poetry in the measurement of an ethereal undulation as in that of the river; for the intellect, during the acts of measurement and calculation, destroys those notions of size which appeal to the poetic sense.   It is a mistake to suppose, with Dr. Young, that

An undevout astronomer is mad ;

there being no necessary connexion between a devout state of mind and the observations and calculations of a practical astronomer.   It is not until the man withdraws from his calculation, as a painter from his work, and thus realizes the great idea on which he has been engaged, that imagination and wonder are excited. There is, I admit, a possible danger here.   If the arithmetical processes of science be too exclusively pursued, they may impair the imagination, and thus the study of Physics is open to the same objection as philological, theological, or political studies, when carried to excess. But even in this case, the injury done is to the investigator himself: it does not reach the mass of mankind.   Indeed, the conceptions furnished by his cold unimaginative reckonings may furnish themes for the poet, and excite in the highest degree that sentiment of wonder which, notwithstanding all its foolish vagaries, table-turning included, I, for my part, should be sorry to see banished from the world.

I have thus far dwelt upon the study of Physics as an agent of intellectual culture; but like other things in Nature, this study subserves more than a single end. The colours of the clouds delight the eye, and, no doubt, accomplish moral purposes also, but the selfsame clouds hold within their fleeces the moisture by which our fields are rendered fruitful.   The sunbeams excite our interest and invite our investigation; but they also extend their beneficent influences to our fruits and corn, and thus accomplish, not only intellectual ends, but minister, at the same time, to our material necessities.   And so it is with scientific research.

While the love of science is a sufficient incentive to the pursuit of science, and the investigator, in the prosecution of his enquiries, is raised above all material considerations, the results of his labours may exercise a potent influence upon the physical condition of the community. This is the arrangement of Nature, and not that of the scientific investigator himself; for he usually pursues his object without regard to its practical applications.

And let him who is dazzled by such applications—who sees in the steam-engine and the electric telegraph the highest embodiment of human genius and the only legitimate object of scientific research, beware of prescribing conditions to the investigator. Let him beware of attempting to substitute for that simple love with which the votary of science pursues his task, the calculations of what he is pleased to call utility. The professed utilitarian is unfortunately, in most cases, the very last man to see the occult sources from which useful results are derived. He admires the flower, but is ignorant of the conditions of its growth. The scientific man must approach Nature in his own way; for if you invade his freedom by your so-called practical considerations, it may be at the expense of those qualities on which his success as a discoverer depends. Let the self-styled practical man look to those from the fecundity of whose thought he, and thousands like him, have sprung into existence. Were they inspired in their first enquiries by the calculations of utility? Not one of them. They were often forced to live low and lie hard, and to seek compensation for their penury in the delight which their favourite pursuits afforded them. In the words of one well qualified to speak upon this subject, ' I say not merely look at the pittance of men like John Dalton, or the voluntary

20

starvation of the late Graff; but compare what is considered as competency or affluence by your Faradays, Liebigs, and Herschels, with the expected results of a life of successful commercial enterprise: then compare the amount of mind put forth, the work done for society in either case, and you will be constrained to allow that the former belong to a class of workers who, properly speaking, are not paid, and cannot be paid for their work, as indeed it is of a sort to which no payment could stimulate.'

But while the scientific investigator, standing upon the frontiers of human knowledge, and aiming at the conquest of fresh soil from the surrounding region of the unknown, makes the discovery of truth his exclusive object for the time, he cannot but feel the deepest interest in the practical application of the truth discovered. There is something ennobling in the triumph of Mind over Matter. Apart even from its uses to society, there is something elevating in the idea of Man having tamed that wild force which flashes through the telegraphic wire, and made it the minister of his will. Our attainments in these directions appear to be commensurate with our needs. We had already subdued horse and mule, and obtained from them all the service which it was in their power to render: we must either stand still, or find more potent agents to execute our purposes. At this point the steam-engine appears. These are still new things; it is not long since we struck into the scientific methods which have produced these results. We cannot for an instant regard them as the final achievements of Science, but rather as an earnest of what she is yet to do. They mark our first great advances upon the dominion of Nature. Animal strength fails, but here are the forces which hold the world together, and the instincts and successes of Man assure him that these forces are his when he is wise enough to command them.

As an instrument of intellectual culture, the study of Physics is profitable to all : as bearing upon special functions, its value, though not so great, is still more tangible. Why, for example, should Members of Parliament be ignorant of the subjects concerning which they are called upon to legislate? In this land of practical physics, why should they be unable to form an independent opinion upon a physical question? Why should the member of a parliamentary committee be left at the mercy of interested disputants when a scientific question is discussed, until he deems the nap a blessing which rescues him from the bewilderments of the committee-room? The education which does not supply the want here referred to, fails in its duty to England. With regard to our working people, in the ordinary sense of the term working, the study of Physics would, I imagine, be profitable, not only as a means of intellectual culture, but also as a moral influence to woo them from pursuits which now degrade them. A man's reformation oftener depends upon the indirect, than upon the direct action of the will. The will must be exerted in the choice of employment which shall break the force of temptation by erecting a barrier against it. The drunkard, for example, is in a perilous condition if he content himself merely with saying, or swearing, that he will avoid strong drink. His thoughts, if not attracted by another force, will revert to the public-house, and to rescue him permanently from this, you must give him an equivalent.

By investing the objects of hourly intercourse with an interest which prompts reflection, new enjoyments would be opened to the working man, and every one of these would be a point of force to protect him against

temptation. Besides this, our factories and our foun-
dries present an extensive field of observation, and
were those who work in them rendered capable, by
previous culture, of *observing* what they *see*, the
results might be incalculable. Who can say what
intellectual Samsons are at the present moment toiling
with closed eyes in the mills and forges of Manchester
and Birmingham? Grant these Samsons sight, and
you multiply the chances of discovery, and with them
the prospects of national advancement. In our multi-
tudinous technical operations we are constantly playing
with forces our ignorance of which is often the cause of
our destruction. There are agencies at work in a
locomotive of which the maker of it probably never
dreamed, but which nevertheless may be sufficient
to convert it into an engine of death. When we
reflect on the intellectual condition of the people who
work in our coal mines, those terrific explosions which
occur from time to time need not astonish us. If these
men possessed sufficient physical knowledge, from the
operatives themselves would probably emanate a system
by which these shocking accidents might be avoided.
Possessed of the knowledge, their personal interests
would furnish the necessary stimulus to its practical
application, and thus two ends would be served at the
same time—the elevation of the men and the diminution of the calamity.

Before the present Course of Lectures was publicly
announced, I had many misgivings as to the propriety
of my taking a part in them, thinking that my place
might be better filled by an older and more experienced
man. To my experience, however, such as it was, I
resolved to adhere, and I have therefore described
things as they revealed themselves to my own eyes, and
have been enacted in my own limited practice. There

is one mind common to us all; and the true expression of this mind, even in small particulars, will attest itself by the response which it calls forth in the convictions of my hearers. I ask your permission to proceed a little further in this fashion, and to refer to a fact or two in addition to those already cited, which presented themselves to my notice during my brief career as a teacher in the college already alluded to. The facts, though extremely humble, and deviating in some slight degree from the strict subject of the present discourse, may yet serve to illustrate an educational principle.

One of the duties which fell to my share was the instruction of a class in mathematics, and I usually found that Euclid and the ancient geometry generally, when properly and sympathetically addressed to the understanding, formed a most attractive study for youth. But it was my habitual practice to withdraw the boys from the routine of the book, and to appeal to their self-power in the treatment of questions not comprehended in that routine. At first, the change from the beaten track usually excited aversion: the youth felt like a child amid strangers; but in no single instance did this feeling continue. When utterly disheartened, I have encouraged the boy by the anecdote of Newton, where he attributes the difference between him and other men, mainly to his own patience; or of Mirabeau, when he ordered his servant, who had stated something to be impossible, never again to use that blockhead of a word. Thus cheered, the boy has returned to his task with a smile, which perhaps had something of doubt in it, but which, nevertheless, evinced a resolution to try again. I have seen his eye brighten, and, at length, with a pleasure of which the ecstasy of Archimedes was but a simple expansion, heard him exclaim, ' I have it, sir.' The consciousness of self-power, thus awakened, was of immense value; and, animated by it, the progress of the class was astonishing. It was often my custom to give the boys the choice of pursuing their propositions in the book, or of trying their strength at others not to be found there. Never in a single instance was the book chosen. I was ever ready to assist when help was needful, but my offers of assistance were habitually declined. The boys had tasted the sweets of intellectual conquest and demanded victories of their own. Their diagrams were scratched on the walls, cut into the beams upon the playground, and numberless other illustrations were afforded of the living interest they took in the subject. For my own part, as far as experience in teaching goes, I was a mere fledgling--knowing nothing of the rules of pedagogics, as the Germans name it; but adhering to the spirit indicated at the commencement of this discourse, and endeavouring to make geometry a means rather than a branch of education. The experiment was successful, and some of the most delightful hours of my existence have been spent in marking the vigorous and cheerful expansion of mental power, when appealed to in the manner here described.

Our pleasure was enhanced when we applied our mathematical knowledge to the solution of physical problems. Many objects of hourly contact had thus a new interest and significance imparted to them. The swing, the see-saw, the tension of the giant-stride ropes, the fall and rebound of the football, the advantage of a small boy over a large one when turning short, particularly in slippy weather; all became subjects of investigation. A lady stands before a looking-glass, of her own height; it was required to know how much of

the glass was really useful to her? We learned with
pleasure the economic fact that she might dispense
with the lower half and see her whole figure notwithstanding. It was also pleasant to prove by mathematics, and verify by experiment, that the angular
velocity of a reflected beam is twice that of the mirror
which reflects it. From the hum of a bee we were
able to determine the number of times the insect flaps
its wings in a second. Following up our researches
upon the pendulum, we learned how Colonel Sabine
had made it the means of determining the figure of
the earth; and we were also startled by the inference
which the pendulum enabled us to draw, that if the
diurnal velocity of the earth were seventeen times its
present amount, the centrifugal force at the equator
would be precisely equal to the force of gravitation, so
that an inhabitant of those regions would then have
the same tendency to fall upwards as downwards. All
these things were sources of wonder and delight to us:
and when we remembered that we were gifted with the
powers which had reached such results, and that the
same great field was ours to work in, our hopes arose
that at some future day we might possibly push the
subject a little further, and add our own victories to
the conquests already won.

I ought to apologise to you for dwelling so long
upon this subject; but the days spent among these
young philosophers made a deep impression on me. I
learned among them something of myself and of human
nature, and obtained some notion of a teacher's vocation.
If there be one profession in England of paramount
importance, I believe it to be that of the schoolmaster;
and if there be a position where selfishness and incompetence do most serious mischief, by lowering the
moral tone and exciting irreverence and cunning where

reverence and noble truthfulness ought to be the
feelings evoked, it is that of the principal of a school.
When a man of enlarged heart and mind comes among
boys,—when he allows his spirit to stream through
them, and observes the operation of his own character
evidenced in the elevation of theirs,—it would be idle
to talk of the position of such a man being honourable.
It is a blessed position. The man is a blessing to
himself and to all around him. Such men, I believe,
are to be found in England, and it behoves those who
busy themselves with the mechanics of education at
the present day, to seek them out. For no matter
what means of culture may be chosen, whether physical
or philological, success must ever mainly depend upon
the amount of life, love, and earnestness, which the
teacher himself brings with him to his vocation.

Let me again, and finally, remind you that the claims
of that science which finds in me to-day its unripened
advocate, are those of the logic of Nature upon the reason
of her child—that its disciplines, as an agent of culture,
are based upon the natural relations subsisting between
Man and the universe of which he forms a part. On
the one side, we have the apparently lawless shifting of
phenomena; on the other side, mind, which requires
law for its equilibrium, and through its own indestructible instincts, as well as through the teachings of
experience, knows that these phenomena are reducible
to law. To chasten this apparent chaos is a problem
which man has set before him. The world was built in
order: and to us are trusted the will and power to
discern its harmonies, and to make them the lessons of
our lives. From the cradle to the grave we are surrounded with objects which provoke inquiry. Descending for a moment from this high plea to considerations
which lie closer to us as a nation—as a land of gas

and furnaces, of steam and electricity : as a land which science, practically applied, has made great in peace and mighty in war :—I ask you whether this 'land of old and just renown ' has not a right to expect from her institutions a culture more in accordance with her present needs than that supplied by declension and conjugation ? And if the tendency should be to lower the estimate of science, by regarding it exclusively as the instrument of material prosperity, let it be the high mission of our universities to furnish the proper counterpoise by pointing out its nobler uses—lifting the national mind to the contemplation of it as the last development of that ' increasing purpose ' which runs through the ages and widens the thoughts of men.

XII.

ON CRYSTALLINE AND SLATY CLEAVAGE.[1]

WHEN the student of physical science has to investigate the character of any natural force, his first
care must be to purify it from the mixture of other forces,
and thus study its simple action. If, for example, he
wishes to know how a mass of liquid would shape itself if
at liberty to follow the bent of its own molecular forces,
he must see that these forces have free and undisturbed
exercise. We might perhaps refer him to the dew-
drop for a solution of the question; but here we have
to do, not only with the action of the molecules of
the liquid upon each other, but also with the action of
gravity upon the mass, which pulls the drop downwards
and elongates it. If he would examine the problem in
its purity, he must do as Plateau has done, detach the
liquid mass from the action of gravity; he would then
find the shape to be a perfect sphere. Natural processes
come to us in a mixed manner, and to the uninstructed
mind are a mass of unintelligible confusion. Suppose
half-a-dozen of the best musical performers to be placed
in the same room, each playing his own instrument to
perfection, but no two playing the same tune; though
each individual instrument might be a source of perfect
music, still the mixture of all would produce mere noise.

[1] From a discourse delivered in the Royal Institution of Great
Britain, June 6, 1856.

Thus it is with the processes of nature, where mechanical and molecular laws intermingle and create apparent confusion. Their mixture constitutes what may
be called the *noise* of natural laws, and it is the vocation
of the man of science to resolve this noise into its com-
ponents, and thus to detect the underlying music.

The necessity of this detachment of one force from
all other forces is nowhere more strikingly exhibited
than in the phenomena of crystallisation. Here, for
example, is a solution of common sulphate of soda or
Glauber salt. Looking into it mentally, we see the
molecules of that liquid, like disciplined squadrons
under a governing eye, arranging themselves into bat-
talions, gathering round distinct centres, and forming
themselves into solid masses, which after a time assume
the visible shape of the crystal now held in my hand.
I may, like an ignorant meddler wishing to hasten
matters, introduce confusion into this order. This may
be done by plunging a glass rod into the vessel; the
consequent action is not the pure expression of the crystalline forces; the molecules rush together with the
confusion of an unorganised mob, and not with the steady
accuracy of a disciplined host. In this mass of bismuth
also we have an example of confused crystallisation; but
in the crucible behind me a slower process is going on:
here there is an architect at work ' who makes no chips,
no din,' and who is now building the particles into
crystals, similar in shape and structure to those beautiful masses which we see upon the table. By permitting
alum to crystallise in this slow way, we obtain these
perfect octahedrons; by allowing carbonate of lime to
crystallise, nature produces these beautiful rhomboids;
when silica crystallises, we have formed these hexagonal
prisms capped at the ends by pyramids; by allowing
saltpetre to crystallise we have these prismatic masses,

and when carbon crystallises, we have the diamond. If we wish to obtain a perfect crystal we must allow the molecular forces free play ; if the crystallising mass be permitted to rest upon a surface it will be flattened, and to prevent this a small crystal must be so suspended as to be surrounded on all sides by the liquid, or, if it rest upon the surface, it must be turned daily so as to present all its faces in succession to the working builder.

In building up crystals these little atomic bricks often arrange themselves into layers which are perfectly parallel to each other, and which can be separated by mechanical means ; this is called the cleavage of the crystal. The crystal of sugar I hold in my hand has thus far escaped the solvent and abrading forces which sooner or later determine the fate of sugar-candy. I readily discover that it cleaves with peculiar facility in one direction. Again I lay my knife upon this piece of rocksalt, and with a blow cleave it in one direction. Laying the knife at right angles to its former position, the crystal cleaves again ; and finally placing the knife at right angles to the two former positions, we find a third cleavage. Rocksalt cleaves in three directions, and the resulting solid is this perfect cube, which may be broken up into any number of smaller cubes. Iceland spar also cleaves in three directions, not at right angles, but oblique to each other, the resulting solid being a rhomboid. In each of these cases the mass cleaves with equal facility in all three directions. For the sake of completeness I may say that many crystals cleave with unequal facility in different directions : heavy spar presents an example of this kind of cleavage.

Turn we now to the consideration of some other phenomena to which the term cleavage may be applied. Beech, deal, and other woods cleave with facility along the fibre, and this cleavage is most perfect when the edge of the axe is laid across the rings which mark the growth of the tree. If you look at this bundle of hay severed from a rick, you will see a sort of cleavage in it also; the stalks lie in horizontal planes, and only a small force is required to separate them laterally. But we cannot regard the cleavage of the tree as the same in character as that of the hayrick. In the one case it is the molecules arranging themselves according to organic laws which produce a cleavable structure, in the other case the easy separation in one direction is due to the mechanical arrangement of the coarse sensible stalks of hay.

This sandstone rock was once a powder held in mechanical suspension by water. The powder was composed of two distinct parts, fine grains of sand and small plates of mica. Imagine a wide strand covered by a tide, or an estuary with water which holds such powder in suspension: how will it sink? The rounded grains of sand will reach the bottom first, because they encounter least resistance, the mica afterwards, and when the tide recedes we have the little plates shining like spangles upon the surface of the sand. Each successive tide brings its charge of mixed powder, deposits its duplex layer day after day, and finally masses of immense thickness are piled up, which by preserving the alternations of sand and mica tell the tale of their formation. Take the sand and mica, mix them together in water, and allow them to subside; they will arrange themselves in the manner indicated, and by repeating the process you can actually build up a mass which shall be the exact counterpart of that presented by nature. Now this structure cleaves with readiness along the planes in which the particles of mica are strewn. Specimens of such a rock sent to me from Halifax, and other masses from the quarries of

Over Darwen in Lancashire, are here before you. With a hammer and chisel I can cleave them into flags; indeed these flags are employed for roofing purposes in the districts from which the specimens have come, and receive the name of slatestone.' But you will discern without a word from me, that this cleavage is not a crystalline cleavage any more than that of a hayrick is. It is molar, not molecular.

This, so far as I am aware of, has never been imagined, and it has been agreed among geologists not to call such splitting as this cleavage at all, but to restrict the term to a phenomenon of a totally different character.

Those who have visited the slate quarries of Cumberland and North Wales will have witnessed the phenomenon to which I refer. We have long drawn our supply of roofing-slates from such quarries; school-boys ciphered on these slates, they were used for tombstones in churchyards, and for billiard-tables in the metropolis; but not until a comparatively late period did men begin to enquire how their wonderful structure was produced. What is the agency which enables us to split Honister Crag, or the cliffs of Snowdon, into laminæ from crown to base? This question is at the present moment one of the great difficulties of geologists, and occupies their attention perhaps more than any other. You may wonder at this. Looking into the quarry of Penrhyn, you may be disposed to offer the explanation I heard given two years ago. 'These planes of cleavage,' said a friend who stood beside me on the quarry's edge, 'are the planes of stratification which have been lifted by some convulsion into an almost vertical position.' But this was a mistake, and indeed here lies the grand difficulty of the problem. The planes of cleavage stand in most cases at a high angle to the bedding. Thanks to Sir Roderick Murchison, I am able to place the proof of this before you. Here is a specimen of slate in which both the planes of cleavage and of bedding are distinctly marked, one of them making a large angle with the other. This is common. The cleavage of slates then is not a question of stratification; what then is its cause?

In an able and elaborate essay published in 1835, Prof. Sedgwick proposed the theory that cleavage is due to the action of crystalline or polar forces subsequent to the consolidation of the rock. 'We may affirm,' he says, 'that no retreat of the parts, no contraction of dimensions in passing to a solid state, can explain such phenomena. They appear to me only resolvable on the supposition that crystalline or polar forces acted upon the whole mass simultaneously in one direction and with adequate force.' And again, in another place: 'Crystalline forces have re-arranged whole mountain masses, producing a beautiful crystalline cleavage, passing alike through all the strata.'[1] The utterance of such a man struck deep, as it ought to do, into the minds of geologists, and at the present day there are few who do not entertain this view either in whole or in part.[2] The boldness of the theory, indeed, has, in some cases, caused speculation to run riot, and we have books published on

[1] *Transactions of the Geological Society*, ser. ii. vol. iii. p. 477.
[2] In a letter to Sir Charles Lyell, dated from the Cape of Good Hope February 20, 1836, Sir John Herschel writes as follows:—'If rocks have been so heated as to allow of a commencement of crystallisation, that is to say, if they have been heated to a point at which the particles can begin to move amongst themselves, or at least on their own axes, some general law must then determine the position in which these particles will rest on cooling. Probably that position will have some relation to the direction in which the heat escapes. Now when all or a majority of particles of the same nature have a general tendency to one position, that must of course determine a cleavage plane.'

the action of polar forces and geologic magnetism, which rather astonish those who know something about the subject. According to this theory whole districts of North Wales and Cumberland, mountains included, are neither more nor less than the parts of a gigantic crystal. These masses of slate were originally fine mud, composed of the broken and abraded particles of older rocks. They contain silica, alumina, potash, soda, and mica mixed mechanically together. In the course of ages the mixture became consolidated, and the theory before us assumes that a process of crystallisation afterwards rearranged the particles and developed in it a single plane of cleavage. Though a bold, and I think inadmissible, stretch of analogies, this hypothesis has done good service. Right or wrong, a thoughtfully uttered theory has a dynamic power which operates against intellectual stagnation; and even by provoking opposition is eventually of service to the cause of truth. It would, however, have been remarkable if, among the ranks of geologists themselves, men were not found to seek an explanation of slate-cleavage involving a less hardy assumption.

The first step in an enquiry of this kind is to seek facts. This has been done, and the labours of Daniel Sharpe (the late President of the Geological Society, who, to the loss of science and the sorrow of all who knew him, has so suddenly been taken away from us), Mr. Henry Clifton Sorby, and others, have furnished us with a body of facts associated with slaty cleavage, and having a most important bearing upon the question.

Fossil shells are found in these slate-rocks. I have here several specimens of such shells in the actual rock, and occupying various positions in regard to the cleavage planes. They are squeezed, distorted, and crushed; in all cases the distortion leads to the inference that the rock which contains these shells has been subjected to enormous pressure in a direction at right angles to the planes of cleavage. The shells are all flattened and spread out in these planes. Compare this fossil trilobite of normal proportions· with these others which have suffered distortion. Some have lain across, some along, and some oblique to the cleavage of the slate in which they are found ; but in all cases the distortion is such as required for its production a compressing force acting at right angles to the planes of cleavage. As the trilobites lay in the mud, the jaws of a gigantic vice appear to have closed upon them and squeezed them into the shapes you see.

We sometimes find a thin layer of coarse gritty material, between two layers of finer rock, through which and across the gritty layer pass the planes of lamination. The coarse layer is found bent by the pressure into sinuosities like a contorted ribbon. Mr. Sorby has described a striking case of this kind. This crumpling can be experimentally imitated; the amount of compression might, moreover, be roughly estimated by supposing the contorted bed to be stretched out, its length measured and compared with the shorter distance into which it has been squeezed. We find in this way that the yielding of the mass has been considerable.

Let me now direct your attention to another proof' of pressure; you see the varying colours which indicate the bedding on this mass of slate. The dark portion is gritty, being composed of comparatively coarse particles, which, owing to their size, shape and gravity, sink first and constitute the bottom of each layer. Gradually, from bottom to top the coarseness diminishes, and near the upper surface we have a layer of exceedingly fine grain. It is the fine mud thus consolidated from which are derived the German razor-stones, so much prized for the sharpening of surgical instruments.

21

When a bed is thin, the fine-grain slate is permitted to rest upon a slab of the coarse slate in contact with it; when the fine bed is thick, it is cut into slices which are cemented to pieces of ordinary slate, and thus rendered stronger. The mud thus deposited is, as might be expected, often rolled up into nodular masses, carried forward, and deposited among coarser material by the rivers from which the slate-mud has subsided. Here are such nodules enclosed in sandstone. Everybody, moreover, who has ciphered upon a school-slate must remember the whitish-green spots which sometimes dotted the surface of the slate, and over which the pencil usually slid as if the spots were greasy. Now these spots are composed of the finer mud, and they could not, on account of their fineness, *bite* the pencil like the surrounding gritty portions of the slate. Here is a beautiful example of these spots : you observe them, on the cleavage surface, in broad round patches. But turn the slate edgeways and the section of each nodule is seen to be a sharp oval with its longer axis parallel to the cleavage. This instructive fact has been adduced by Mr. Sorby. I have made excursions to the quarries of Wales and Cumberland, and to many of the slate yards of London, and found the fact general. Thus we elevate a common experience of our boyhood into evidence of the highest significance as regards a most important geological problem. From the magnetic deportment of these slates, I was led to infer that these spots contain a less amount of iron than the surrounding dark slate. An analysis was made for me by Mr. Hambly in the laboratory of Dr. Percy at the School of Mines with the following result :—

ANALYSIS OF SLATE.

Dark Slate, two analyses.

1. Percentage of iron	5·85
2. ,, ,, ...	6·13
	Mean . . 5·99

Whitish Green Slate.

1. Percentage of iron	3·24
2. ,, ,, ...	3·12
	Mean . . 3·18

According to these analyses the quantity of iron in the dark slate immediately adjacent to the greenish spot is nearly double the quantity contained in the spot itself. This is about the proportion which the magnetic experiments suggested.

Let me now remind you that the facts brought before you are typical—each is the representative of a class. We have seen shells crushed; the trilobites squeezed, beds contorted, nodules of greenish marl flattened; and all these sources of independent testimony point to one and the same conclusion, namely, that slate-rocks have been subjected to enormous pressure in a direction at right angles to the planes of cleavage.

In reference to Mr. Sorby's contorted bed, I have said that by supposing it to be stretched out and its length measured, it would give us an idea of the amount of yielding of the mass above and below the bed. Such a measurement, however, would not give the exact amount of yielding. I hold in my hand a specimen of slate with its bedding marked upon it; the lower portions of each layer being composed of a comparatively coarse gritty material something like what you may suppose the contorted bed to be composed of. Now in crossing these gritty portions, the cleavage turns, as if tending to cross the bedding at another angle. When the pressure began to act, the intermediate bed, which is not entirely unyielding, suffered longitudinal pressure; as it bent, the pressure became gradually more transverse, and the direction of its cleavage is exactly such as you would infer from an action of this kind—it is neither quite across the bed, nor yet in the same direction as the cleavage of the slate above and below it, but intermediate between both. Supposing the cleavage to be at right angles to the pressure, this is the direction which it ought to take across these more unyielding strata.

Thus we have established the concurrence of the phenomena of cleavage and pressure—that they accompany each other; but the question still remains, Is the pressure sufficient to account for the cleavage? A single geologist, as far as I am aware, answers boldly in the affirmative. This geologist is Sorby, who has attacked the question in the true spirit of a physical investigator. Call to mind the cleavage of the flags of Halifax and Over Darwen, which is caused by the interposition of layers of mica between the gritty strata. Mr. Sorby finds plates of mica to be also a constituent of slate-rock. He asks himself, what will be the effect of pressure upon a mass containing such plates confusedly mixed up in it? It will be, he argues, and he argues rightly, to place the plates with their flat surfaces more or less perpendicular to the direction in which the pressure is exerted. He takes scales of the oxide of iron, mixes them with a fine powder, and on squeezing the mass finds that the tendency of the scales is to set themselves at right angles to the line of pressure. Along the planes of weakness produced by the scales the mass cleaves.

By tests of a different character from those applied

by Mr. Sorby, it might be shown how true his conclusion is—that the effect of pressure on elongated
particles, or plates, will be such as he describes it.
But while the scales must be regarded as a true cause,
I should not ascribe to them a large share in the production of the cleavage. I believe that even if the
plates of mica were wholly absent, the cleavage of
slate-rocks would be much the same as it is at present.

Here is a mass of pure white wax : it contains no
mica particles, no scales of iron, or anything analogous
to them. Here is the selfsame substance submitted to
pressure. I would invite the attention of the eminent
geologists now before me to the structure of this wax.
No slate ever exhibited so clean a cleavage ; it splits
into laminæ of surpassing tenuity, and proves at a
single stroke that pressure is sufficient to produce
cleavage, and that this cleavage is independent of inter-
mixed plates or scales. I have purposely mixed this
wax with elongated particles, and am unable to say at
the present moment that the cleavage is sensibly
affected by their presence—if anything, I should say
they rather impair its fineness and clearness than pro-
mote it.

The finer the slate is the more perfect will be the
resemblance of its cleavage to that of the wax. Compare the surface of the wax with the surface of this
slate from Borrodale in Cumberland. You have precisely the same features in both : you see flakes clinging
to the surfaces of each, which have been partially torn
away in cleaving. Let any close observer compare
these two effects, he will, I am persuaded, be led to
the conclusion that they are the product of a common
cause.[1]

_____

[1] I have usually softened the wax by warming it, kneaded it
with the fingers, and pressed it between thick plates of glass pre-

But you will ask me how, according to my view, does pressure produce this remarkable result? This may be stated in a very few words.

There is no such thing in nature as a body of perfectly homogeneous structure. I break this clay which seems so uniform, and find that the fracture presents to my eyes innumerable surfaces along which it has given way, and it has yielded along those surfaces because in them the cohesion of the mass is less than elsewhere. I break this marble, and even this wax, and observe the same result; look at the mud at the bottom of a dried pond; look at some of the ungravelled walks in Kensington Gardens on drying after rain,— they are cracked and split, and other circumstances being equal, they crack and split where the cohesion is a minimum. Take then a mass of partially consolidated mud. Such a mass is divided and subdivided by interior surfaces along which the cohesion is comparatively small. Penetrate the mass in idea, and you will see it composed of numberless irregular polyhedra bounded by surfaces of weak cohesion. Imagine such a mass subjected to pressure,—it yields and spreads out in the direction of least resistance;[1] the little polyhedra become converted into laminæ, separated from each other by surfaces of weak cohesion, and the infallible result

viously wetted. At the ordinary summer temperature the pressed wax is soft, and tears rather than cleaves; on this account I cool my compressed specimens in a mixture of pounded ice and salt, and when thus cooled they split cleanly.

[1] It is scarcely necessary to say that if the mass were squeezed equally in all directions no laminated structure could be produced; it must have room to yield in a lateral direction. Mr. Warren De la Rue informs me that he once wished to obtain white-lead in a fine granular state, and to accomplish this he first compressed it. The mould was conical, and permitted the lead to spread out a little laterally. The lamination was as perfect as that of slate, and it quite defeated him in his effort to obtain a granular powder.

will be a tendency to cleave at right angles to the line
of pressure.

Further, a mass of dried mud is full of cavities and
fissures. If you break dried pipe-clay you see them in
great numbers, and there are multitudes of them so
small that you cannot see them. A flattening of these
cavities must take place in squeezed mud, and this
must to some extent facilitate the cleavage of the mass
in the direction indicated.

Although the time at my disposal has not permitted
me duly to develope these thoughts, yet for the last
twelve months the subject has presented itself to me
almost daily under one aspect or another. I have
never eaten a biscuit during this period without remarking the cleavage developed by the rolling-pin.
You have only to break a biscuit across, and to look at
the fracture, to see the laminated structure. We have
here the means of pushing the analogy further. I invite you to compare the structure of this slate, which
was subjected to a high temperature during the conflagration of Mr. Scott Russell's premises, with that of a
biscuit. Air or vapour within the slate has caused it
to swell, and the mechanical structure it reveals is precisely that of a biscuit. During these enquiries I have
received much instruction in the manufacture of puff-
paste. Here is some such paste baked under my own
superintendence. The cleavage of our hills is acci-
dental cleavage, but this is cleavage with intention.
The volition of the pastrycook has entered into its
formation. It has been his aim to preserve a series of
surfaces of structural weakness, along which the dough
divides into layers. Puff-paste in preparation must
not be handled too much; it ought, moreover, to be
rolled on a cold slab, to prevent the butter from melt-

ing, and diffusing itself, thus rendering the paste more homogeneous and less liable to split. Puff-paste is, then, simply an exaggerated case of slaty cleavage.

The principle here enunciated is so simple as to be almost trivial; nevertheless, it embraces not only the cases mentioned, but, if time permitted, it might be shown you that the principle has a much wider range of application. When iron is taken from the puddling furnace it is more or less spongy, an aggregate n fact of small nodules: it is at a welding heat, and at this temperature is submitted to the process of rolling. Bright smooth bars are the result. But notwithstanding the high heat the nodules do not perfectly blend together. The process of rolling draws them into fibres. Here is a mass acted upon by dilute sulphuric acid, which exhibits in a striking manner this fibrous structure. The experiment was made by my friend Dr. Percy, without any reference to the question of cleavage.

Break a piece of ordinary iron and you have a granular fracture; beat the iron, you elongate these granules, and finally render the mass fibrous. Here are pieces of rails along which the wheels of locomotives have slidden; the granules have yielded and become plates. They exfoliate or come off in leaves; all these effects belong, I believe, to the great class of phenomena of which slaty cleavage forms the most prominent example.[1]

We have now reached the termination of our task. You have witnessed the phenomena of crystallisation, and have had placed before you the facts which are found associated with the cleavage of slate rocks. Such facts, as expressed by Helmholtz, are so many telescopes to our spiritual vision, by which we can see backward

[1] For some further observations on this subject by Mr. Sorby and myself, see *Philosophical Magazine* for August, 1856.

through the night of antiquity, and discern the forces
which have been in operation upon the earth's surface

> Ere the lion roared,
> Or the eagle soared.

From evidence of the most independent and
trustworthy character, we come to the conclusion
that these slaty masses have been subjected to enormous
pressure, and by the sure method of experiment we
have shown—and this is the only really new point
which has been brought before you—how the pressure
is sufficient to produce the cleavage. Expanding our
field of view, we find the self-same law, whose footsteps we
trace amid the crags of Wales and Cumberland, extending into the domain of the pastrycook and ironfounder;
nay, a wheel cannot roll over the half-dried mud of our
streets without revealing to us more or less of the features
of this law.   Let me say, in conclusion, that the spirit
in which this problem has been attacked by geologists,
indicates the dawning of a new day for their science.
The great intellects who have laboured at geology, and
who have raised it to its present pitch of grandeur, were
compelled to deal with the subject in mass; they had
no time to look after details.   But the desire for more
exact knowledge is increasing; facts are flowing in
which, while they leave untouched the intrinsic wonders
of geology, are gradually supplanting by solid truths the
uncertain speculations which beset the subject in its in-
fancy.   Geologists now aim to imitate, as far as possible,
the conditions of nature, and to produce her results;
they are approaching more and more to the domain of
physics, and I trust the day will soon come when we
shall interlace our friendly arms across the common
boundary of our sciences, and pursue our respective tasks

in a spirit of mutual helpfulness, encouragement and goodwill.

[I would now lay more stress on the lateral yielding, referred to in the note at the bottom of page 316, accompanied as it is by tangential sliding, than I was prepared to do when this lecture was given. This sliding is, I think, the principal cause of the planes of weakness, both in pressed wax and slate rock.   J. T. 1871.]

XIII.

ON PARAMAGNETIC AND DIAMAGNETIC FORCES.[1]

THE notion of an attractive force, which draws bodies towards the centre of the earth, was entertained by Anaxagoras and his pupils, by Democritus, Pythagoras, and Epicurus; and the conjectures of these ancients were renewed by Galileo, Huyghens, and others, who stated that bodies attract each other as a magnet attracts iron. Kepler applied the notion to bodies beyond the surface of the earth, and affirmed the extension of this force to the most distant stars. Thus it would appear, that in the attraction of iron by a magnet originated the conception of the force of gravitation. Nevertheless, if we look closely at the matter, it will be seen that the magnetic force possesses characters strikingly distinct from those of the force which holds the universe together. The theory of gravitation is, that every particle of matter attracts every other particle; in magnetism also we have attraction, but we have always, at the same time, repulsion, the final effect being due to the difference of these two forces. A body may be intensely acted on by a magnet, and still no motion of translation will follow, if the repulsion be equal to the attraction. Previous to magnetization, a dipping needle, when its centre of gravity is supported, stands accurately level; but, after magnetization, one end of it, in our latitude,

[1] Abstract of a discourse delivered in the Royal Institution, February 1, 1856.

is pulled towards the north pole of the earth. The needle, however, being suspended from the arm of a fine balance, its weight is found unaltered by its magnetization. In like manner, when the needle is permitted to float upon a liquid, and thus to follow the attraction of the north magnetic pole of the earth, there is no motion of the mass towards that pole. The reason is known to be, that although the marked end of the needle is attracted by the north pole, the unmarked end is repelled by an equal force, the two equal and opposite forces neutralizing each other.

When the pole of an ordinary magnet is brought to act upon the swimming needle, the latter is attracted,—the reason being that the attracted end of the needle being nearer to the pole of the magnet than the repelled end, the force of attraction is the more powerful of the two. In the case of the earth, its pole is so distant that the length of the needle is practically zero. In like manner, when a piece of iron is presented to a magnet, the nearer parts are attracted, while the more distant parts are repelled; and because the attracted portions are nearer to the magnet than the repelled ones, we have a balance in favour of attraction. Here then is the special characteristic of the magnetic force, which distinguishes it from that of gravitation. The latter is a simple unpolar force, while the former is duplex or polar. Were gravitation like magnetism, a stone would no more fall to the ground than a piece of iron towards the north magnetic pole: and thus, however rich in consequences the supposition of Kepler and others may have been, it is clear that a force like that of magnetism would not be able to transact the business of the universe.

The object of this discourse is to enquire whether the force of diamagnetism, which manifests itself as a repulsion of certain bodies by the poles of a magnet, is to be ranged as a polar force, beside that of magnetism ; or as an unpolar force, beside that of gravitation. When a cylinder of soft iron is placed within a wire helix, and surrounded by an electric current, the antithesis of its two ends, or, in other words, its polar excitation, is at once manifested by its action upon a magnetic needle ; and it may be asked why a cylinder of bismuth may not be substituted for the cylinder of iron, and its state similarly examined. The reason is, that the excitement of the bismuth is so feeble, that it would be quite masked by that of the helix in which it is enclosed; and the problem that now meets us is, so to excite a diamagnetic body that the pure action of the body upon a magnetic needle may be observed, unmixed with the action of the body used to excite the diamagnetic.

How this has been effected may be illustrated in the following manner :—When through an upright helix of covered copper wire, a voltaic current is sent, the top of the helix attracts, while its bottom repels, the same pole of a magnetic needle ; its central point, on the contrary, is neutral, and exhibits neither attraction nor repulsion. Such a helix is caused to stand between the

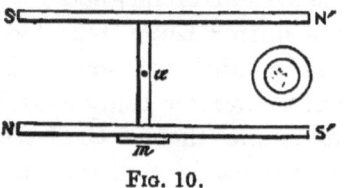

FIG. 10.

two poles N's' of an astatic system.[1] The two magnets s N' and s'N are united by a rigid cross piece at

---

[1] The reversal of the poles of the two magnets, which were of the same strength, completely annulled the action of the earth as a magnet.

their centres, and are suspended from the point a, so that both magnets swing in the same horizontal plane. It is so arranged that the poles N's' are opposite to the central or neutral point of the helix, so that when a current is sent through the latter, the magnets, as before explained, are unaffected. Here then we have an excited helix which itself has no action upon the magnets, and we are thus enabled to examine the action of a body placed within the helix and excited by it, undisturbed by the influence of the latter. The helix being 12 inches high, a cylinder of soft iron 6 inches long, suspended from a string and passing over a pulley, can be raised or lowered within the helix. When it is so far sunk that its lower end rests upon the table, the upper end finds itself between the poles N's' of the astatic system. The iron cylinder is thus converted into a strong magnet, attracting one of the poles, and repelling the other, and consequently deflecting the entire astatic system. When the cylinder is raised so that the upper end is at the level of the top of the helix, its lower end comes between the poles N's'; and a deflection opposed in direction to the former one is the immediate consequence. To render these deflections more easily visible, a mirror m is attached to the system of magnets; a beam of light thrown upon the mirror being reflected and projected as a bright disk against the wall. The distance of this image from the mirror being considerable, and its angular motion double that of the latter, a very slight motion of the magnet is sufficient to produce a displacement of the image through several yards.

This then is the principle of the beautiful apparatus[1]

[1] Devised by Prof. W. Weber, and constructed by M. Leyser, of Leipzig.

by which the investigation was conducted. It is manifest that if a second helix be placed between the poles
s n with a cylinder within it, the action upon the
astatic magnet may be exalted. This was the arrangement made use of in the actual enquiry. Thus to
intensify the feeble action, which it is here our object
to seek, we have in the first place neutralized the action
of the earth upon the magnets, by placing them asta-
tically. Secondly, by making use of two cylinders, and
permitting them to act simultaneously on the four poles
of the magnets, we have rendered the deflecting force
four times what it would be, if only a single pole
were used. Finally, the whole apparatus was enclosed in a suitable case which protected the magnets
from air-currents, and the deflections were read off
through a glass plate in the case, by means of a telescope and scale placed at a considerable distance from
the instrument.

A pair of bismuth cylinders was first examined.
Sending a current through the helices, and observing
that the magnets swung perfectly free, it was first arranged that the bismuth cylinders within the helices
had their central or neutral points opposite to the poles
of the magnets. All being at rest the number on the
scale marked by the cross wire of the telescope was 572.
The cylinders were then moved, one up the other down,
so that two of their ends were brought to bear simultaneously upon the magnetic poles : the magnet moved
promptly, and after some oscillations[1] came to rest at
the number 612; thus moving from a smaller to a
larger number. The other two ends of the bars were
next brought to bear upon the magnet: a prompt deflection was the consequence, and the final position of

[1] To lessen these a copper damper was made use of.

equilibrium was 526 : the movement being from a larger
to a smaller number.   We thus observe a manifest polar
action of the bismuth cylinders upon the magnet ; one
pair of ends deflecting it in one direction, and the other
pair deflecting it in the opposite direction.

Substituting for the cylinders of bismuth thin
cylinders of iron, of magnetic slate, of sulphate of iron,
carbonate of iron, protochloride of iron, red ferrocyanide
of potassium, and other magnetic bodies, it was found
that when the position of the magnetic cylinders was
the same as that of the cylinders of bismuth, the deflection produced by the former was always opposed in
direction to that produced by the latter ; and hence the
disposition of the force in the diamagnetic body must
have been precisely antithetical to its disposition in the
magnetic ones.

But it will be urged, and indeed has been urged
against this inference, that the deflection produced by
the bismuth cylinders may be due to induced currents
excited in the metal by its motion within the helices.
In reply to this objection, it may be stated, in the
first place, that the deflection is permanent, and cannot therefore be due to induced currents, which are only
of momentary duration.   It has also been urged that
such experiments ought to be made with other metals,
and with better conductors than bismuth ; for if due
to currents of induction, the better the conductor the
more exalted will be the effect.   This requirement was
complied with.

Cylinders of antimony were substituted for those of
bismuth.   This metal is a better conductor of electricity, but less strongly diamagnetic than bismuth.
If therefore the action referred to be due to induced
currents we ought to have it greater in the case of antimony than with bismuth ; but if it springs from a true

diamagnetic polarity, the action of the bismuth ought to exceed that of the antimony. Experiment proves this to be the case. Hence the deflection produced by these metals is due to their diamagnetic, and not to their conductive capacity. Copper cylinders were next examined : here we have a metal which conducts electricity fifty times better than bismuth, but its diamagnetic power is nearly null; if the effects be due to induced currents we ought to have them here in an enormously exaggerated degree, but no sensible deflection was produced by the two cylinders of copper.

It has also been proposed by the opponents of diamagnetic polarity to coat fragments of bismuth with some insulating substance, so as to render the formation of induced currents impossible, and to test the question with cylinders of these fragments. This requirement was also fulfilled. It is only necessary to reduce the bismuth to powder and expose it for a short time to the air to cause the particles to become so far oxidised as to render them perfectly insulating. The insulating power of the powder was exhibited experimentally; nevertheless, this powder, enclosed in glass tubes, exhibited an action scarcely less powerful than that of the massive bismuth cylinders.

But the most rigid proof, a proof admitted to be conclusive by those who have denied the antithesis of magnetism and diamagnetism, remains to be stated. Prisms of the same heavy glass as that with which the diamagnetic force was discovered, were substituted for the metallic cylinders, and their action upon the magnet was proved to be precisely the same in kind as that of the cylinders of bismuth. The enquiry was also extended to other insulators : to phosphorus, sulphur, nitre, calcareous spar, statuary marble, with the same invariable result : each of these substances was proved

22.

to be polar, the disposition of the force being the same as that of bismuth and the reverse of that of iron.   When a bar of iron is set erect, its lower end is known to be a north pole, and its upper end a south pole, in virtue of the earth's induction.   A marble statue, on the contrary, has its feet a south pole, and its head a north pole, and there is no doubt that the same remark applies to its living archetype; each man walking over the earth's surface is a true diamagnet, with its poles the reverse of those of a mass of magnetic matter of the same shape and position.

An experiment of practical value, as affording a ready estimate of the different conductive powers of two metals for electricity, was exhibited in the lecture, for the purpose of proving experimentally some of the statements made in reference to this subject.   A cube of bismuth was suspended by a twisted string between the two poles of an electro-magnet.   The cube was attached by a short copper wire to a little square pyramid, the base of which was horizontal, and its sides formed of four small triangular pieces of looking-glass. A beam of light was suffered to fall upon this reflector, and as the reflector followed the motion of the cube the images cast from its sides followed each other in succession, each describing a circle about thirty feet in diameter.   As the velocity of rotation augmented, these images blended into a continuous ring of light. At a particular instant the electro-magnet was excited, currents were evolved in the rotating cube, and the strength of these currents, which increases with the conductivity of the cube for electricity, was practically estimated by the time required to bring the cube and its associated mirrors to a state of rest.   With bismuth this time amounted to a score of seconds or more: a cube of copper, on the contrary, was struck almost instantly motionless when the circuit was established.

XIV

PHYSICAL BASIS OF SOLAR CHEMISTRY.[1]

OMITTING all preface, attention was first drawn to an experimental arrangement intended to prove that gaseous bodies radiate heat in different degrees. Near a double screen of polished tin was placed an ordinary ring gas-burner, and on this was placed a hot copper ball, from which a column of heated air ascended. Behind the screen, but so situated that no ray from the ball could reach the instrument, was an excellent thermo-electric pile, connected by wires with a very delicate galvanometer. The pile was known to be an instrument whereby heat is applied to the generation of electric currents; the strength of the current being an accurate measure of the quantity of the heat. As long as both faces of the pile are at the same temperature, no current is produced; but the slightest difference in the temperature of the two faces at once declares itself by the production of a current, which, when carried through the galvanometer, indicates by the deflection of the needle both its strength and its direction.

The two faces of the pile were in the first instance brought to the same temperature; the equilibrium being shown by the needle of the galvanometer standing

[1] From a discourse delivered at the Royal Institution of Great Britain, June 7, 1861.

at zero. The rays emitted by the current of hot air already referred to were permitted to fall upon one of the faces of the pile; and an extremely slight movement of the needle showed that the radiation from the hot air, though sensible, was extremely feeble. Connected with the ring-burner was a holder containing oxygen gas; and by turning a cock, a stream of this gas was permitted to issue from the burner, strike the copper ball, and ascend in a heated column in front of the pile. The result was, that oxygen showed itself, as a radiator of heat, to be quite as feeble as atmospheric air.

A second holder containing olefiant gas was then connected with the ring-burner. Oxygen and air had already flowed over the ball and cooled it in some degree. Hence the olefiant gas laboured under a disadvantage. But on permitting the gas to rise from the ball, it casts an amount of heat against the adjacent face of the pile sufficient to impel the needle of the galvanometer almost to 90°. This experiment proved the vast difference between two equally invisible gases with regard to their power of emitting radiant heat.

The converse experiment was now performed. The thermo-electric pile was removed and placed between two cubes filled with water kept in a state of constant ebullition; and it was so arranged that the quantities of heat falling from the cubes on the opposite faces of the pile were exactly equal, thus neutralising each other. The needle of the galvanometer being at zero, a sheet of oxygen gas was caused to issue from a slit between one of the cubes and the adjacent face of the pile. If this sheet of gas possessed any sensible power of intercepting the thermal rays from the cube, one face of the pile being deprived of the heat thus intercepted, a difference of temperature between its two faces would instantly set in, and the result would be declared by the galvanometer. The quantity absorbed by the oxygen under those circumstances was too feeble to affect the galvanometer; the gas, in fact, proved perfectly transparent to the rays of heat. It had but a feeble power of radiation: it had an equally feeble power of absorption.

The pile remaining in its position, a sheet of olefiant gas was caused to issue from the same slit as that through which the oxygen had passed. No one present could see the gas; it was quite invisible, the light went through it as freely as through oxygen or air; but its effect upon the thermal rays emanating from the cube was what might be expected from a sheet of metal. A quantity so large was cut off, that the needle of the galvanometer, promptly quitting the zero line, moved with energy to its stops. Thus the olefiant gas, so light and clear and pervious to luminous rays, was proved to be a most potent destroyer of the rays emanating from an obscure source. The reciprocity of action established in the case of oxygen comes out here; the good radiator is found by this experiment to be the good absorber.

This result, now exhibited before a public audience for the first time, was typical of what had been obtained with gases generally. Going through the entire list of gases and vapours in this way, we find radiation and absorption to be as rigidly associated as positive and negative in electricity, or as north and south polarity in magnetism. So that if we make the number which expresses the absorptive power the numerator of a fraction, and that which expresses its radiative power the denominator, the result would be, that on account of the numerator and denominator varying in the same proportion, the value of that fraction would always remain the same, whatever might be the gas or vapour experimented with.

But why should this reciprocity exist? What is the meaning of absorption? what is the meaning of radiation? When you cast a stone into still water, rings of waves surround the place where it falls; motion is radiated on all sides from the centre of disturbance. When a hammer strikes a bell, the latter vibrates; and sound, which is nothing more than an undulatory motion of the air, is radiated in all directions. Modern philosophy reduces light and heat to the same mechanical category. A luminous body is one with its atoms in a state of vibration; a hot body is one with its atoms also vibrating, but at a rate which is incompetent to excite the sense of vision ; and, as a sounding body has the air around it, through which it propagates its vibrations, so also the luminous or heated body has a medium, called ether, which accepts its motions and carries them forward with inconceivable velocity. Radiation, then, as regards both light and heat, is the transference of motion from the vibrating body to the ether in which it swings: and, as in the case of sound, the motion imparted to the air is soon transferred to surrounding objects, against which the aërial undulations strike, the sound being, in technical language, *absorbed*; so also with regard to light and heat, absorption consists in the transference of motion from the agitated ether to the molecules of the absorbing body.

The simple atoms are found to be bad radiators; the compound atoms good ones: and the higher the degree of complexity in the atomic grouping, the more potent, as a general rule, is the radiation and absorption. Let us get definite ideas here, however gross, and purify them afterwards by the process of abstraction. Imagine our simple atoms swinging like single spheres in the ether; they cannot create the swell which a group of them united to form a system can produce. An oar runs freely edgeways through the water, and imparts far less of its motion to the water than when its broad flat side is brought to bear upon it. In our present language the oar, broad side vertical, is a good radiator ; broad side horizontal, it is a bad radiator. Conversely the waves of water, impinging upon the flat face of the oar-blade, will impart a greater amount of motion to it than when impinging upon the edge. In the position in which the oar radiates well, it also absorbs well. Simple atoms glide through the ether without much resistance; compound ones encounter resistance, and hence yield up more speedily their motion to the ether. *Mix* oxygen and nitrogen mechanically, they absorb and radiate a certain amount of heat. Cause these gases to *combine* chemically and form nitrous oxide, both the absorption and radiation are thereby augmented hundreds of times !

In this way we look with the telescope of the in-tellect into atomic systems, and obtain a conception of processes which the eye of sense can never reach. But gases and vapours possess a power of choice as to the rays which they absorb. They single out certain groups of rays for destruction, and allow other groups to pass unharmed. This is best illustrated by a famous experi-ment of Sir David Brewster's, modified to suit present requirements. Into a glass cylinder, with its ends stopped by discs of plate-glass, a small quantity of nitrous acid gas is introduced ; the presence of the gas being indicated by its rich brown colour. The beam from an electric lamp being sent through two prisms of bisulphide of carbon, a spectrum seven feet long and eighteen inches wide is cast upon the screen. Introducing the cylinder containing the nitrous acid into the path of the beam as it issues from the lamp, the splendid and continuous spectrum becomes instantly furrowed by numerous dark bands, the rays answering to which are intercepted by the nitric gas, while the light which falls upon the intervening spaces is permitted to pass with comparative impunity.

Here also the principle of reciprocity, as regards radiation and absorption, holds good ; and could we, without otherwise altering its physical character, render that nitrous gas luminous, we should find that the very rays which it absorbs are precisely those which it would emit. When atmospheric air and other gases are brought to a state of intense incandescence by the passage of an electric spark, the spectra which we obtain from them consist of a series of bright bands. But such spectra are produced with the greatest brilliancy when, instead of ordinary gases, we make use of metals heated so highly as to volatilise them. This is easily done by the voltaic current. A capsule of carbon filled with mercury, which formed the positive electrode of the electric lamp, has a carbon point brought down upon it. On separating the one from the other, a brilliant arc containing the mercury in a volatilised condition passes between them. The spectrum of this arc is not continuous like that of the solid carbon points, but consists of a series of vivid bands, each corresponding in colour to that particular portion of the spectrum to which its rays belong. Copper gives its system of bands; zinc gives its system; and brass, which is an alloy of copper and zinc, gives a spectrum made up of the bands belonging to both metals.

Not only, however, when metals are united like zinc and copper to form an alloy, is it possible to obtain the bands which belong to them. No matter how we may disguise the metal— allowing it to unite with oxygen to form an oxide, and this again with an acid to form a salt; if the heat applied be sufficiently intense, the bands belonging to the metal reveal themselves with

perfect definition.   Into holes drilled in a cylinder of
retort carbon, pure culinary salt is introduced.   When
the carbon is made the positive electrode of the
lamp, the resultant spectrum shows the brilliant yellow
lines of the metal sodium.   Similar experiments made
with the chlorides of strontium, calcium, lithium,[1] and
other metals, give the bands due to the respective
metals.   When different salts are mixed together, and
rammed into holes in the carbon; a spectrum is obtained which contains the bands of them all.

The position of these bright bands never varies, and
each metal has its own system.   Hence the competent
observer can infer from the bands of the spectrum the
metals which produce it.   It is a language addressed to
the eye instead of the ear; and the certainty would not
be augmented if each metal possessed the power of
audibly calling out, ' I am here ! '   Nor is this language
affected by distance.   If we find that the sun or the
stars give us the bands of our terrestrial metals, it is a
declaration on the part of these orbs that such metals
enter into their composition.   Does the sun give us any
such intimation ?   Does the solar spectrum exhibit bright
lines which we might compare with those produced by
our terrestrial metals, and prove either their identity
or difference ?   No.   The solar spectrum, when closely
examined, gives us a multitude of fine dark lines instead
of bright ones.   They were first noticed by Dr. Wollaston, but were multiplied and investigated with profound
skill by Fraunhofer, and named after him Fraunhofer's

---

[1] The vividness of the colours of the lithium spectrum is extraordinary; the spectrum, moreover, contained a blue band of indescribable splendour.   It was thought by many, during the
discourse, that I had mistaken strontium for lithium, as this blue
band had never before been seen.   I have obtained it many times
since; and my friend Dr. Miller, having kindly analysed the substance made use of, pronounces it pure chloride of lithium.—J. T.

lines.   They had been long a standing puzzle to philosophers.   The bright lines yielded by metallic vapours
had been also known to us for years ; but the connection between both classes of phenomena was wholly
unknown, until Kirchhoff, with admirable acuteness,
revealed the secret, and placed it at the same time in
our power to chemically analyse the sun.

We have now some difficult work before us.   Hitherto we have been delighted by objects which addressed
themselves as much to our æsthetic taste as to our scientific faculty ; we have ridden pleasantly to the base of
the final cone of Etna, and must now dismount and
march through ashes and lava, if we would enjoy the
prospect from the summit.   Our problem is to connect
the dark lines of Fraunhofer with the bright ones of the
metals.   The white beam of the lamp is refracted in
passing through our two prisms, but its different com-
ponents are refracted in different degrees, and thus its
colours are drawn apart.   Now the colour depends solely
upon the rate of oscillation of the atoms of the luminous body ; red light being produced by one rate, blue .
light by a much quicker rate, and the colours between
red and blue by the intermediate rates.   The solid incandescent coal-points give us a continuous spectrum ;
or in other words they emit rays of all possible periods
between the two extremes of the spectrum.   Colour, as
many of you know, is to light what *pitch* is to sound.
When a violin-player presses his finger on a string he
makes it shorter and tighter, and thus, causing it to
vibrate more speedily, heightens the pitch.   Imagine
such a player to move his fingers slowly along the string,
shortening it gradually as he draws his bow, the note
would rise in pitch by a regular gradation ; there would
be no gap intervening between note and note.   Here
we have the analogue to the continuous spectrum, whose
colours insensibly blend together without gap or inter-

ruption, from the red of the lowest pitch to the violet of
the highest.   But suppose the player, instead of gradually shortening his string, to press his finger on a certain point, and to sound the corresponding note ; then
to pass on to another point more or less distant, and
sound its note ; then to another, and so on, thus sounding
particular notes separated from each other by gaps which
correspond to the intervals of the string passed over ; we
should then have the exact analogue of a spectrum composed of separate bright bands with intervals of darkness
between them.   But this, though a perfectly true and intelligible analogy, is not sufficient for our purpose ; we
must look with the mind's eye at the oscillating atoms of
the volatilised metal.   Figure these atoms as connected
together by springs of a certain tension, which, if the
atoms are squeezed together, push them again asunder,
and if the atoms are drawn apart, pull them again
together, causing them, before coming to rest, to quiver
for a certain time at a certain definite rate determined
by the strength of the spring.   Now the volatilised metal
. which gives us one bright band is to be figured as having
its atoms united by springs all of the same tension, its
vibrations are all of one kind.   The metal which gives
us two bands may be figured as having some of its atoms
united by springs of one tension, and others by springs
of a different tension.   Its vibrations are of two distinct
kinds ; so also when we have three or more bands we are
to figure as many distinct sets of springs, each capable
of vibrating in its own particular time and at a different
rate from the others.   If we seize this idea definitely, we
shall have no difficulty in dropping the metaphor of
springs, and substituting for it mentally the forces by
which the atoms act upon each other.   Having thus far
cleared our way, let us make another effort to advance.

A heavy ivory ball is here suspended from a string,
I blow against this ball; a single puff of my breath

moves it a little way from its position of rest; it swings
back towards me, and when it reaches the limit of its
swing I puff again.   It now swings further; and thus by
timing the puffs I can so accumulate their action as to
produce oscillations of large amplitude.   The ivory ball
here has absorbed the motion which my breath communicated to the air.   I now bring the ball to rest.
Suppose, instead of the breath, a wave of air to strike
against it, and that this wave is followed by a series of
others which succeed each other exactly in the same
intervals as my puffs; it is obvious that these waves
would communicate their motion to the ball and cause
it to swing as the puffs did.   And it is equally manifest
that this would not be the case if the impulses of the
waves were not properly timed; for then the motion
imparted to the pendulum by one wave would be neutralised by another, and there could not be the accumulation of effect obtained when the periods of the waves
correspond with the periods of the pendulum.   So much
for the particular impulses absorbed by the pendulum.
But if such a pendulum set oscillating in air could pro-.
duce waves in the air, it is evident that the waves it
would produce would be of the same period as those
whose motions it would take up or absorb most completely, if they struck against it.

Perhaps the most curious effect of these timed
impulses ever described was that observed by a watch-
maker, named Ellicott, in the year 1741.   He left two
clocks leaning against the same rail; one of them,
which we may call A, was set going; the other, B, not.
Some time afterwards he found, to his surprise, that B
was ticking also.   The pendulums being of the same
length, the shocks imparted by the ticking of A to the
rail against which both clocks rested were propagated
to B, and were so timed as to set B going.   Other

curious effects were at the same time observed.   When
the pendulums differed from each other a certain amount,
A set B going, but the reaction of B stopped A.   Then
B set A going, and the reaction of A stopped B.   When
the periods of oscillation were close to each other, but
still not quite alike, the clocks mutually controlled
each other, and by a kind of compromise they ticked
in perfect unison.

But what has all this to do with our present subject?
The varied actions of the universe are all modes of
motion; and the vibration of a ray claims strict brotherhood with the vibrations of our pendulum.   Suppose
ethereal waves striking upon atoms which oscillate in
the same periods as the waves, the motion of the waves
will be absorbed by the atoms; suppose we send our
beam of white light through a sodium flame, the atoms
of that flame will be chiefly affected by those undulations which are synchronous with their own periods of
vibration.   There will be on the part of those particular
rays a transference of motion from the agitated ether
to the atoms of the volatilised metal, which, as already
defined, is absorption.

The experiment justifying this conclusion is now for
the first time to be made before a public audience.   I pass
a beam through our two prisms, and the spectrum spreads
its colours upon the screen.   Between the lamp and the
prism I interpose a snapdragon light.   Alcohol and
water are here mixed with common salt, and the metal
dish that holds them is heated by a spirit-lamp.   The
vapour from the mixture ignites and we have a monochromatic flame.   Through this flame the beam from the
lamp is now passing; and observe the result upon the
spectrum.   You see a shady band cut out of the yellow,
—not very dark, but sufficiently so to be seen by everybody present.

But let me exalt this effect. Placing in front of the electric lamp the intense flame of a large Bunsen's burner, a platinum capsule containing a bit of sodium less than a pea in magnitude is plunged into the flame. The sodium soon volatilises and burns with brilliant incandescence. The beam crosses the flame, and at the same time the yellow band of the spectrum is clearly and sharply cut out, a band of intense darkness occupying its place. On withdrawing the sodium, the brilliant yellow of the spectrum takes its proper place, while the reintroduction of the flame causes the band to reappear.

Let me be more precise :—The yellow colour of the spectrum extends over a sensible space, blending on one side with the orange and on the other with the green. The term ' yellow band ' is therefore somewhat indefinite. This vagueness may be entirely removed. By dipping the carbon-point used for the positive electrode into a solution of common salt, and replacing it in the lamp, the bright yellow band produced by the sodium vapour stands out from the spectrum. When the sodium flame is caused to act upon the beam it is that particular yellow band that is obliterated, an intensely black streak occupying its place.

An additional step of reasoning leads to the conclusion that if, instead of the flame of sodium alone, we were to introduce into the path of the beam a flame in which lithium, strontium, magnesium, calcium, &c., are in a state of volatilisation, each metallic vapour would cut out a system of bands, corresponding exactly in position with the bright bands of the same metallic vapour. The light of our electric lamp shining through such a composite flame would give us a spectrum cut up by dark lines, exactly as the solar spectrum is cut up by the lines of Fraunhofer.

Thus by the combination of the strictest reasoning

with the most conclusive experiment, we reach the solution of one of the grandest of scientific problems— the constitution of the sun.   The sun consists of a nucleus surrounded by a flaming atmosphere.   The light of the nucleus would give us a continuous spectrum, like that of our common carbon-points ; but having to pass through the photosphere, as our beam had to pass through the flame, those rays of the nucleus which the photosphere can itself emit are absorbed, and shaded spaces, corresponding to the particular rays absorbed, occur in the spectrum.   Abolish the solar nucleus, and we should have a spectrum showing a bright line in the place of every dark line of Fraunhofer.   These lines are therefore not absolutely dark, but dark by an amount corresponding to the difference between the light of the nucleus intercepted by the photosphere, and the light which issues from the latter.

The man to whom we owe this noble generalisation is Kirchhoff, Professor of Natural Philosophy in the University of Heidelberg ; [1] but, like every other great discovery, it is compounded of various elements.   Mr. Talbot observed the bright lines in the spectra of coloured flames.   Sixteen years ago Dr. Miller gave drawings and descriptions of the spectra of various coloured flames.   Wheatstone, with his accustomed ingenuity, analysed the light of the electric spark, and showed that the metals between which the spark passed determined the bright bands in the spectrum of the spark.   Masson published a prize essay on these bands; Van der Willigen, and more recently Plücker, have given us beautiful drawings of the spectra, obtained from the discharge of Ruhmkorff's coil.   But none of these distinguished men betrayed the least knowledge of the connection between the bright bands of the

[1] Now Professor in the University of Berlin.

metals and the dark lines of the solar spectrum. The man who came nearest to the philosophy of the subject was Ångström. In a paper translated from Poggendorff's 'Annalen' by myself, and published in the 'Philosophical Magazine' for 1855, he indicates that the rays which a body absorbs are precisely those which it can emit when rendered luminous. In another place, he speaks of one of his spectra giving the general impression of a *reversal* of the solar spectrum. Foucault, Stokes, and Thomson, have all been very close to the discovery; and, for my own part, the examination of the radiation and absorption of heat by gases and vapours, some of the results of which I placed before you at the commencement of this discourse, would have led me in 1859 to the law on which all Kirchhoff's speculations are founded, had not an accident withdrawn me from the investigation. But Kirchhoff's claims are unaffected by these circumstances. True, much that I have referred to formed the necessary basis of his discovery; so did the laws of Kepler furnish to Newton the basis of the theory of gravitation. But what Kirchhoff has done carries us far beyond all that had before been accomplished. He has introduced the order of law amid a vast assemblage of empirical observations, and has ennobled our previous knowledge by showing its relationship to some of the most sublime of natural phenomena.

XV.

ELEMENTARY MAGNETISM.

A LECTURE TO SCHOOLMASTERS.

WE have no reason to believe that the sheep or the dog, or indeed any of the lower animals, feel an interest in the laws by which natural phenomena are regulated. A herd may be terrified by a thunderstorm; birds may go to roost, and cattle return to their stalls, during a solar eclipse; but neither birds nor cattle, as far as we know, ever think of enquiring into the causes of these things. It is otherwise with man. The presence of natural objects, the occurrence of natural events, the varied appearances of the universe in which he dwells penetrate beyond his organs of sense, and appeal to an inner power of which the senses are the mere instruments and excitants. No fact is to him either original or final. He cannot limit himself to the contemplation of it alone, but endeavours to ascertain its position in a series to which uniform experience assures him it must belong. He regards all that he witnesses in the present as the efflux and sequence of something that has gone before, and as the source of a system of events which is to follow. The notion of spontaneity, by which in his ruder state he accounted for natural events, is abandoned; the idea that nature is an aggregate of independent parts also disappears, as the connection and mutual dependence of physical

23

powers become more and more manifest: until he is finally led to regard Nature as an organic whole—as a body each of whose members sympathises with the rest, changing, it is true, from age to age, but changing without break of continuity in the relation of cause and effect.

The system of things which we call Nature is, however, too vast and various to be studied first-hand by any single mind. As knowledge extends there is always a tendency to subdivide the field of investigation. Its various parts are taken up by different minds, and thus receive a greater amount of attention than could possibly be bestowed on them if each investigator aimed at the mastery of the whole. The centrifugal form in which knowledge, as a whole, advances, spreading ever wider on all sides, is due in reality to the exertions of individuals, each of whom directs his efforts, more or less, along a single line. Accepting, in many respects, his culture from his fellow-men—taking it from spoken words or from written books—in some one direction, the student of Nature ought actually to *touch* his work. He may otherwise be a distributor of knowledge, but not a creator, and he fails to attain that vitality of thought, and correctness of judgment, which direct and habitual contact with natural truth can alone impart.

One large department of the system of Nature which forms the chief subject of my own studies, and to which it is my duty to call your attention this evening, is that of physics, or natural philosophy. This term is large enough to cover the study of Nature generally, but it is usually restricted to a department which, perhaps, lies closer to our perceptions than any other. It deals with the phenomena and laws of light and heat—with the phenomena and laws of magnetism and electricity—with those of sound—with the pressures and motions of liquids and gases, whether at rest or in a state of translation or of undulation. The science of mechanics is a portion of natural philosophy, though at present so large as to need the exclusive attention of him who would cultivate it profoundly. Astronomy is the application of physics to the motions of the heavenly bodies, the vastness of the field causing it, however, to be regarded as a department in itself. In chemistry physical agents play important parts. By heat and light we cause atoms and molecules to unite or to fall asunder. Electricity exerts a similar power. Through their ability to separate nutritive compounds into their constituents, the solar beams build up the whole vegetable world, and by it the animal world. The touch of the self-same beams causes hydrogen and chlorine to unite with sudden explosion, and to form by their combination a powerful acid. Thus physics and chemistry intermingle. Physical agents are, however, employed by the chemist as a means to an end; while in physics proper the laws and phenomena of the agents themselves, both qualitative and quantitative, are the primary objects of attention.

My duty here to-night is to spend an hour in telling how this subject is to be studied, and how a knowledge of it is to be imparted to others. From the domain of physics, which would be unmanageable as a whole, I select as a sample the subject of magnetism. I might readily entertain you on the present occasion with an account of what natural philosophy has accomplished. I might point to those applications of science of which we hear so much in the newspapers, and which are so often mistaken for science itself. I might, of course, ring changes on the steam-engine and the telegraph, the electrotype and the photograph, the medical appli-

cations of physics, and the various other inlets by which
scientific thought filters into practical life.   That would
be easy compared with the task of informing you how
you are to make the study of physics the instrument of
your pupil's culture; how you are to possess its facts
and make them living seeds which shall take root and
grow in the mind, and not lie like dead lumber in the
storehouse of memory.   This is a task much heavier
than the mere recounting of scientific achievements;
and it is one which, feeling my own want of time to
execute it aright, I might well hesitate to accept.

But let me sink excuses, and attack the work before
me.   First and foremost, then, I would advise you to
get a knowledge of facts from actual observation.
Facts looked at directly are vital; when they pass into
words half the sap is taken out of them.   You wish,
for example, to get a knowledge of magnetism; well,
provide yourself with a good book on the subject, if
you can, but do not be content with what the book tells
you; do not be satisfied with its descriptive woodcuts;
see the operations of the force yourself.   Half of our
book writers describe experiments which they never
made, and their descriptions often lack both force and
truth ; but, no matter how clever or conscientious they
may be, their written words cannot supply the place of
actual observation.   Every fact has numerous radiations, which are shorn off by the man who describes it.
Go, then, to a philosophical instrument maker, and
give a shilling or half a crown for a straight bar-
magnet, or, if you can afford it, purchase a pair of
them; or get a smith to cut a length of ten inches
from a bar of steel an inch wide and half an inch
thick; file its ends smoothly, harden it, and get somebody like myself to magnetise it.   Procure some darning needles, and also a little unspun silk, which will

give you a suspending fibre void of torsion. Make a little loop of paper, or of wire, and attach your fibre to it. Do it neatly. In the loop place a darning-needle, and bring the two ends or poles, as they are called, of your bar-magnet successively up to the ends of the needle. Both the poles, you find, attract both ends of the needle. Replace the needle by a bit of annealed iron wire; the same effects ensue. Suspend successively little rods of lead, copper, silver, brass, wood, glass, ivory, or whalebone; the magnet produces no sensible effect upon any of the substances. You thence infer a special property in the case of steel and iron. Multiply your experiments, however, and you will find that some other substances, besides iron and steel, are acted upon by your magnet. A rod of the metal nickel, or of the metal cobalt, from which the blue colour used by painters is derived, exhibits powers similar to those observed with the iron and steel.

In studying the character of the force you may, however, confine yourself to iron and steel, which are always at hand. Make your experiments with the darning-needle over and over again; operate on both ends of the needle; try both ends of the magnet. Do not think the work dull; you are conversing with Nature, and must acquire over her language a certain grace and mastery, which practice can alone impart. Let every movement be made with care, and avoid slovenliness from the outset. Experiment, as I have said, is the language by which we address Nature, and through which she sends her replies; in the use of this language a lack of straightforwardness is as possible, and as prejudicial, as in the spoken language of the tongue. If, therefore, you wish to become acquainted with the truth of Nature, you must from the first resolve to deal with her sincerely.

Now remove your needle from its loop, and draw it from eye to point along one of the ends of the magnet; resuspend it, and repeat your former experiment. You now find that each extremity of the magnet attracts one end of the needle, and repels the other. The simple attraction observed in the first instance, is now replaced by a *dual* force. Repeat the experiment till you have thoroughly observed the ends which attract and those which repel each other.

Withdraw the magnet entirely from the vicinity of your needle, and leave the latter freely suspended by its fibre. Shelter it as well as you can from currents of air, and if you have iron buttons on your coat, or a steel penknife in your pocket, beware of their action. If you work at night, beware of iron candlesticks, or of brass ones with iron rods inside. Freed from such disturbances, the needle takes up a certain determinate position. It sets its length nearly north and south. Draw it aside and let it go. After several oscillations it will again come to the same position. If you have obtained your magnet from a philosophical instrument maker, you will see a mark on one of its ends. Supposing, then, that you drew your needle along the end thus marked, and that the point of your needle was the last to quit the magnet, you will find that the point turns to the south, the eye of the needle turning towards the north. Make sure of this, and do not take the statement on my authority.

Now take a second darning-needle like the first, and magnetise it in precisely the same manner: freely suspended it also will turn its eye to the north and its point to the south. Your next step is to examine the action of the two needles which you have thus magnetised upon each other.

Take one of them in your hand, and leave the other

suspended; bring the eye-end of the former near the
eye-end of the latter; the 'suspended needle retreats:
it is repelled.   Make the same experiment with the
two points; you obtain the same result, the suspended
needle is repelled.   Now cause the dissimilar ends to
act on each other—you have attraction—point attracts
eye, and eye attracts point.   Prove the reciprocity of
this action by removing the suspended needle, and
putting the other in its place.   You obtain the same
result.   The attraction, then, is mutual, and the repulsion
is mutual.   You have thus demonstrated in the clearest
manner the fundamental law of magnetism, that like
poles repel, and that unlike poles attract, each other.
You may say that this is all easily understood without
doing; but *do it*, and your knowledge will not be confined to what I have uttered here.

I have said that one end of your bar magnet has a
mark upon it; lay several silk fibres together, so as to
get sufficient strength, or employ a thin silk ribbon,
and form a loop large enough to hold your magnet.
Suspend it; it turns its marked end towards the north.
This marked end is that which in England is called the
north pole.   If a common smith has made your magnet,
it will be convenient to determine its north pole yourself,
and to mark it with a file.   Vary your experiments by
causing your magnetised darning-needle to attract and
repel your large magnet; it is quite competent to do
so.   In magnetising the needle, I have supposed the
point to be the last to quit the marked end of the
magnet; the point of the needle is a south pole.   The
end which last quits the magnet is always opposed in
polarity to the end of the magnet with which it has been
last in contact.

You may perhaps learn all this in a single hour; but
spend several at it, if necessary; and remember, under-

standing it is not sufficient: you must obtain a manual aptitude in addressing Nature. If you speak to your fellow-man you are not entitled to use jargon. Bad experiments are jargon addressed to Nature, and just as much to be deprecated. Manual dexterity in illustrating the interaction of magnetic poles is of the utmost importance at this stage of your progress; and you must not neglect attaining this power over your implements. As you proceed, moreover, you will be tempted to do more than I can possibly suggest. Thoughts will occur to you which you will endeavour to follow out: questions will arise which you will try to answer. The same experiment may be twenty different things to twenty people. Having witnessed the action of pole on pole, through the air, you will perhaps try whether the magnetic power is not to be screened off. You use plates of glass, wood, slate, pasteboard, or gutta-percha, but find them all pervious to this wondrous force. One magnetic pole acts upon another through these bodies as if they were not present. Should you ever become a patentee for the regulation of ships' compasses, you will not fall, as some projectors have done, into the error of screening off the magnetism of the ship by the interposition of such substances.

If you wish to teach a class you must contrive that the effects which you have thus far witnessed for yourself shall be witnessed by twenty or thirty pupils. And here your private ingenuity must come into play. You will attach bits of paper to your needles, so as to render their movements visible at a distance, denoting the north and south poles by different colours, say green and red. You may also improve upon your darning-needle. Take a strip of sheet steel, heat it to vivid redness and plunge it into cold water. It is thereby hardened; rendered, in fact, almost as brittle as glass. Six inches of this, magnetised in the manner of the darning-

needle, will be better able to carry your paper indexes.
Having secured such a strip, you proceed thus :—

Magnetise a small sewing-needle and determine its
poles ; or, break half an inch, or an inch, off your magnetised darning-needle and suspend it by a fine silk fibre.
The sewing-needle, or the fragment of the darning
needle, is now to be used as a test-needle, to examine
the distribution of the magnetism in your strip of steel.
Hold the strip upright in your left hand, and cause the
test-needle to approach the lower end of your strip ;
one end of the test-needle is attracted, the other is
repelled.   Raise your needle along the strip ; its oscillations, which at first were quick, become slower ;
opposite the middle of the strip they cease entirely ;
neither end of the needle is attracted ; above the
middle the test-needle turns suddenly round, its other
end being now attracted.   Go through the experiment thoroughly : you thus learn that the entire lower
half of the strip attracts one end of the needle, while
the entire upper half attracts the opposite end.   Supposing the north end of your little needle to be that
attracted below, you infer that the entire lower half of
your magnetised strip exhibits south magnetism, while
the entire upper half exhibits north magnetism.   So
far, then, you have determined the distribution of
magnetism in your strip of steel.

You look at this fact, you think of it ; in its sug-
gestiveness the value of an experiment chiefly consists.
The thought naturally arises : ' What will occur if I
break my strip of steel across in the middle ?   Shall I
obtain two magnets each possessing a single pole ? '
Try the experiment ; break your strip of steel, and test
each half as you tested the whole.   The mere presenta-
tion of its two ends in succession to your test-needle,
suffices to show that you have *not* a magnet with a

single pole—that each half possesses two poles with a
neutral point between them.   And if you again break
the half into two other halves, you will find that each
quarter of the original strip exhibits precisely the same
magnetic distribution as the whole strip.   You may
continue the breaking process: no matter how small
your fragment may be, it still possesses two opposite
poles and a neutral point between them.   Well, your
hand ceases to break where breaking becomes a mechanical impossibility; but does the mind stop there?
No: you follow the breaking process in idea when you
can no longer realise it in fact; your thoughts wander
amid the very atoms of your steel, and you conclude
that each atom is a magnet, and that the force exerted
by the strip of steel is the mere summation, or resultant,
of the forces of its ultimate particles.

Here, then, is an exhibition of power which we can
call forth at pleasure or cause to disappear.   We magnetise our strip of steel by drawing it along the pole of
a magnet; we can demagnetise it, or reverse its magnetism, by properly drawing it along the same pole in
the opposite direction.   What, then, is the real nature
of this wondrous change?   What is it that takes place
among the atoms of the steel when the substance is
magnetised?   The question leads us beyond the region
of sense, and into that of imagination.   This faculty,
indeed, is the divining-rod of the man of science.
Not, however, an imagination which catches its creations from the air, but one informed and inspired by
facts; capable of seizing firmly on a physical image
as a principle, of discerning its consequences, and of
devising means whereby these forecasts of thought
may be brought to an experimental test.   If such a
principle be adequate to account for all the phenomena
—if from an assumed cause the observed acts neces-

sarily follow, we call the assumption a theory, and, once possessing it, we can not only revive at pleasure facts already known, but we can predict others which we have never seen. Thus, then, in the prosecution of physical science, our powers of observation, memory, imagination, and inference, are all drawn upon. We observe facts and store them up ; the constructive imagination broods upon these memories, tries to discern their interdependence and weave them to an organic whole. The theoretic principle flashes or slowly dawns upon the mind; and then the deductive faculty interposes to carry out the principle to its logical consequences. A perfect theory gives dominion over natural facts ; and even an assumption which can only partially stand the test of a comparison with facts, may be of eminent use in enabling us to connect and classify groups of phenomena. The theory of magnetic fluids is of this latter character, and with it we must now make ourselves familiar.

With the view of stamping the thing more firmly on your minds, I will make use of a strong and vivid image. In optics, red and green are called complementary colours; their mixture produces *white*. Now I ask you to imagine each of these colours to possess a self-repulsive power ; that red repels red, that green repels green ; but that red attracts green and green attracts red, the attraction of the dissimilar colours being equal to the repulsion of the similar ones. Imagine the two colours mixed so as to produce white, and suppose two strips of wood painted with this white ; what will be their action upon each other ? Suspend one of them freely as we suspended our darning needle, and bring the other near it ; what will occur ? The red component of the strip you hold in your hand will repel the red component of your suspended strip ; but then it will attract the green, and, the forces being equal, they neutralise each other. In fact, the least reflection shows you that the strips will be as indifferent to each other as two unmagnetised darning-needles would be under the same circumstances.

But suppose, instead of mixing the colours, we painted one half of each strip from centre to end red, and the other half green, it is perfectly manifest that the two strips would now behave towards each other exactly as our two magnetised darning-needles—the red end would repel the red and attract the green, the green would repel the green and attract the red; so that, assuming two colours thus related to each other, we could by their mixture produce the neutrality of an unmagnetised body, while by their separation we could produce the duality of action of magnetised bodies.

But you have already anticipated a defect in my conception; for if we break one of our strips of wood in the middle we have one half entirely red, and the other entirely green, and with these it would be impossible to imitate the action of our broken magnet. How, then, must we modify our conception? We must evidently suppose *each molecule of the wood* painted green on one face and red on the opposite one. The resultant action of all the atoms would then exactly resemble the action of a magnet. Here also, if the two opposite colours of each atom could be caused to mix so as to produce white, we should have, as before, perfect neutrality.

For these two self-repellent and mutually attractive colours, substitute in your minds two invisible self-repellent and mutually attractive fluids, which in ordinary steel are mixed to form a neutral compound, but which the act of magnetisation separates from each other, placing the opposite fluids on the opposite face of each molecule. You have then a perfectly distinct conception of the celebrated theory of magnetic fluids. The strength of the magnetism excited is supposed to be proportional to the quantity of neutral fluid decomposed. According to this theory nothing is actually transferred from the exciting magnet to the excited steel. The act of magnetisation consists in the forcible separation of two fluids which existed in the steel before it was magnetised, but which then neutralised each other by their coalescence. And if you test your magnet, after it has excited a hundred pieces of steel, you will find that it has lost no force—no more, indeed, than I should lose, had my words such a magnetic influence on your minds as to excite in them a strong resolve to study natural philosophy. I should rather be the gainer by my own utterance, and by the reaction of your fervour. The magnet also is the gainer by the reaction of the body which it magnetises.

Look now to your excited piece of steel; figure each molecule with its opposed fluids spread over its opposite faces. How can this state of things be permanent? The fluids, by hypothesis, attract each other; what, then, keeps them apart? Why do they not instantly rush together across the equator of the atom, and thus neutralise each other? To meet this question philosophers have been obliged to infer the existence of a special force, which holds the fluids asunder. They call it *coercive force*; and it is found that those kinds of steel which offer most resistance to being magnetised —which require the greatest amount of 'coercion' to tear their fluids asunder—are the very ones which offer the greatest resistance to the reunion of the fluids, after they have been once separated. Such kinds of steel are most suited to the formation of *permanent* magnets. It is manifest, indeed, that without coercive force a permanent magnet would not be at all possible.

Probably long before this you will have dipped the end of your magnet among iron filings, and observed how they cling to it; or into a nail-box, and found how it drags the nails after it. I know very well that if you are not the slaves of routine, you will have by this time done many things that I have not told you to do, and thus multiplied your experience beyond what I have indicated. You are almost sure to have caused a bit of iron to hang from the end of your magnet, and you have probably succeeded in causing a second bit to attach itself to the first, a third to the second; until finally the force has become too feeble to bear the weight of more. If you have operated with nails, you may have observed that the points and edges hold together with the greatest tenacity; and that a bit of iron clings more firmly to the corner of your magnet than to one of its flat surfaces. In short, you will in all likelihood have enriched your experience in many ways without any special direction from me.

Well, the magnet attracts the nail, and the nail attracts a second one. This proves that the nail in contact with the magnet has had the magnetic quality developed in it by that contact. If it be withdrawn from the magnet its power to attract its fellow nail ceases. Contact, however, is not necessary. A sheet of glass or paper, or a space of air, may exist between the magnet and the nail; the latter is still magnetised, though not so forcibly as when in actual contact. The nail thus presented to the magnet is itself a temporary magnet. That end which is turned towards the magnetic pole has the opposite magnetism of the pole which excites it; the end most remote from the pole has the same magnetism as the pole itself, and between the two poles the nail, like the magnet, possesses a magnetic equator.

Conversant as you now are with the theory of magnetic fluids, you have already, I doubt not, anticipated me in imagining the exact condition of an iron nail under the influence of the magnet. You picture the iron as possessing the neutral fluid in abundance; you picture the magnetic pole, when brought near, decomposing the fluid; repelling the fluid of a like kind with itself, and attracting the unlike fluid; thus exciting in the parts of the iron nearest to itself the opposite polarity. But the iron is incapable of becoming a permanent magnet. It only shows its virtue as long as the magnet acts upon it. What, then, does the iron lack which the steel possesses ? It lacks coercive force. Its fluids are separated with ease; but, once the separating cause is removed, they flow together again, and neutrality is restored. Imagination must be quite nimble in picturing these changes—able to see the fluids dividing and reuniting, according as the magnet is brought near or withdrawn. Fixing a definite pole in your mind, you must picture the precise arrangement of the two fluids with reference to this pole, and be able to arouse similar pictures in the minds of your pupils. You will cause them to place magnets and iron in various positions, and describe the exact magnetic state of the iron in each particular case. The mere facts of magnetism will have their interest immensely augmented by an acquaintance with the principles whereon the facts depend. Still, while you use this theory of magnetic fluids to track out the phenomena and link them together, you will not forget to tell your pupils that it is to be regarded as a symbol merely,—a symbol, moreover, which is incompetent to cover all the facts,[1] but which does good

---

[1] This theory breaks down when applied to diamagnetic bodies

practical service whilst we are waiting for the actual truth.

The state of excitement into which iron is thrown by the influence of a magnet, is sometimes called 'magnetisation by influence.' More commonly, however, the magnetism is said to be 'induced' in the iron, and hence this mode of magnetising is called 'magnetic induction.' Now, there is nothing theoretically perfect in Nature: there is no iron so soft as not to possess a certain amount of coercive force, and no steel so hard as not to be capable, in some degree, of magnetic induction. The quality of steel is in some measure possessed by iron, and the quality of iron is shared in some degree by steel. It is in virtue of this latter fact that the unmagnetised darning-needle was attracted in your first experiment; and from this you may at once deduce the consequence that, after the steel has been magnetised, the repulsive action of a magnet must be always less than its attractive action. For the repulsion is opposed by the inductive action of the magnet on the steel, while the attraction is assisted by the same inductive action. Make this clear to your minds, and verify it by your experiments. In some cases you can actually make the attraction due to the temporary magnetism overbalance the repulsion due to the permanent magnetism, and thus cause two poles of the same kind apparently to attract each other. When, however, good hard magnets act on each other from a sufficient distance, the inductive action practically vanishes, and the repulsion of like poles is sensibly equal to the attraction of unlike ones.

which are repelled by magnets. Like soft iron, such bodies are thrown into a state of temporary excitement, in virtue of which they are repelled; but any attempt to explain such a repulsion by the decomposition of a fluid will demonstrate its own futility.

I dwell thus long on elementary principles, because
they are of the first importance, and it is the temptation
of this age of unhealthy cramming to neglect them.
Now follow me a little farther. In examining the
distribution of magnetism in your strip of steel you
raised the needle slowly from bottom to top, and found
what we called a neutral point at the centre. Now
does the magnet really exert no influence on the pole
presented to its centre? Let us see.

Let s n, fig. 11, be our magnet, and let n represent
a particle of north magnetism placed exactly opposite
the middle of the magnet. Of course this is an ima-
ginary case, as you can never in reality thus detach
your north magnetism from its neighbour. But supposing us to have done so, what would be the action of
the two poles of the magnet on n? Your reply will of
course be that the pole s attracts n while the pole n
repels it. Let the magnitude and direction of the
attraction be expressed by the line $n\,m$, and the mag-

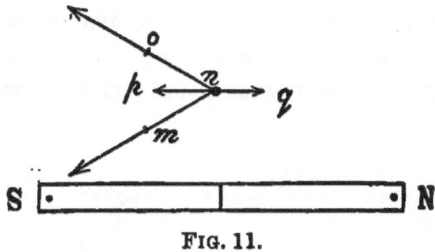

FIG. 11.

nitude and direction of the repulsion by the line $n\,o$.
Now, the particle n being equally distant from s and n,
the line $n\,o$, expressing the repulsion, will be equal to
$m\,n$, which expresses the attraction. Acted upon by
two such forces, the particle n must evidently move in
the direction $n\,p$, exactly midway between $m\,n$ and $n\,o$.
Hence you see that, although there is no tendency of

24

the particle n to move towards the magnetic equator,
there is a tendency on its part to move parallel to the
magnet. If, instead of a particle of north magnetism,
we placed a particle of south magnetism opposite to
the magnetic equator, it would evidently be urged
along the line nq; and if, instead of two separate
particles of magnetism, we place a little magnetic
needle, containing both north and south magnetism,
opposite the magnetic equator, its south pole being
urged along nq, and its north along np, the little
needle will be compelled to set itself parallel to the
magnet s N. Make the experiment, and satisfy yourselves that this is a true deduction.

Substitute for your magnetic needle a bit of iron
wire, devoid of permanent magnetism, and it will set
itself exactly as the needle does. Acted upon by the
magnet, the wire, as you know, becomes a magnet and
behaves as such; it will turn its north pole towards p,
and south pole towards q, just like the needle.

But supposing you shift the position of your particle
of north magnetism, and bring it nearer to one end of
your magnet than to the other; the forces acting on
the particle are no longer equal; the nearest pole of

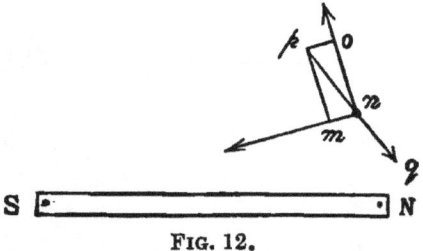

FIG. 12.

the magnet will act more powerfully on the particle
than the more distant one. Let s N, fig. 12, be the
magnet, and n the particle of north magnetism, in its

new position. It is repelled by N, and attracted by s.
Let the repulsion be represented in magnitude and
direction by the line *n o*, and the attraction by the
shorter line *n m*. The resultant of these two forces
will be found by completing the parallelogram *m n o p*,
and drawing its diagonal *n p*. Along *n p*, then, a
particle of north magnetism would be urged by the
simultaneous action of s and N. Substituting a particle
of south magnetism for *n*, the same reasoning would
lead to the conclusion that the particle would be urged
along *n q*. If we place at *n* a short magnetic needle,
its north pole will be urged along *n p*, its south pole
along *n q*, the only position possible to the needle,
thus acted on, being along the line *p q*, which is no
longer parallel to the magnet. Verify this deduction
by actual experiment.

In this way we might go round the entire magnet;
and, considering its two poles as two centres from
which the force emanates, we could, in accordance with
ordinary mechanical principles, assign a definite direction to the magnetic needle at every particular place.
And substituting, as before, a bit of iron wire for the
magnetic needle, the positions of both will be the
same.

Now, I think, without further preface, you will be
able to comprehend for yourselves, and explain to
others, one of the most interesting effects in the whole
domain of magnetism. Iron filings you know are
particles of iron, irregular in shape, being longer in
some directions than in others. For the present experiment, moreover, instead of the iron filings, very
small scraps of thin iron wire might be employed. I
place a sheet of paper over the magnet; it is all the
better if the paper be stretched on a wooden frame, as
this enables us to keep it quite level. I scatter the

FIG. 13.

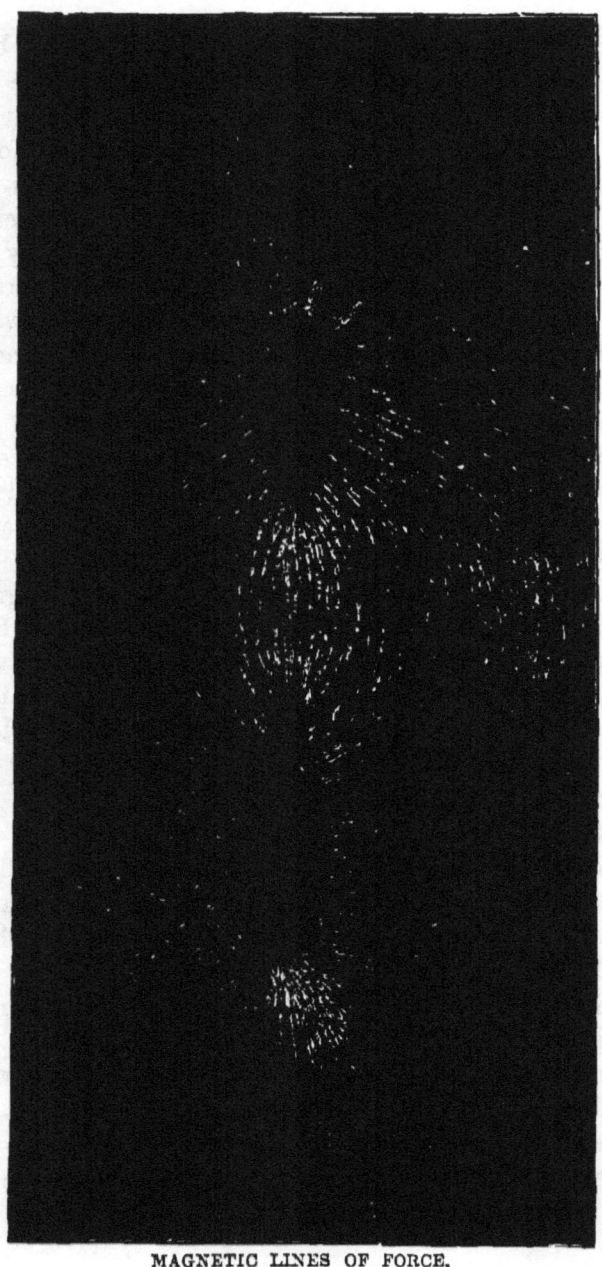

MAGNETIC LINES OF FORCE.
From a Photograph by PROFESSOR MAYER.

filings, or the scraps of wire, from a sieve upon the
paper, and tap the latter gently, so as to liberate the
particles for a moment from its friction. The magnet
acts on the filings through the paper, and see how it
arranges them! They embrace the magnet in a series
of beautiful curves, which are technically called ' magnetic curves,' or ' lines of magnetic force.' Does the
meaning of these lines yet flash upon you? Set your
magnetic needle, or your suspended bit of wire, at any
point of one of the curves, and you will find the direction of the needle, or of the wire, to be exactly that of
the particle of iron, or of the magnetic curve, at that
point. Go round and round the magnet; the direction
of your needle always coincides with the direction of
the curve on which it is placed. These, then, are the
lines along which a particle of south magnetism, if you
could detach it, would move to the north pole, and a
bit of north magnetism to the south pole. They are
the lines along which the decomposition of the neutral
fluid takes place. In the case of the magnetic needle,
one of its poles being urged in one direction, and the
other pole in the opposite direction, the needle must
necessarily set itself as a *tangent* to the curve. I will
not seek to simplify this subject further. If there be
anything obscure or confused or incomplete in my
statement, you ought now, by patient thought, to be
able to clear away the obscurity, to reduce the confusion to order, and to supply what is needed to render
the explanation complete. Do not quit the subject
until you thoroughly understand it; and if you are
then able to look with your mind's eye at the play of
forces around a magnet, and see distinctly the operation
of those forces in the production of the magnetic curves,
the time which we have spent together will not have
been spent in vain.

In this thorough manner we must master our materials, reason upon them, and, by determined study,
attain to clearness of conception.   Facts thus dealt
with exercise an expansive force upon the intellect ;—
they widen the mind to generalisation.   We soon
recognise a brotherhood between the larger phenomena
of Nature and the minute effects which we have observed in our private chambers.   Why, we enquire,
does the magnetic needle set north and south ?   Evidently it is compelled to do so by the earth ; the great
globe which we inherit is itself a magnet.   Let us
learn a little more about it.   By means of a bit of
wax, or otherwise, attach the end of your silk fibre to
the middle point of your magnetic needle ; the needle
will thus be uninterfered with by the paper loop, and
will enjoy to some extent a power of 'dipping' its
point, or its eye, below the horizon.   Lay your bar
magnet on a table, and hold the needle over the equator
of the magnet.   The needle sets horizontal.   Move it
towards the north end of the magnet ; the south end
of the needle dips, the dip augmenting as you approach
the north pole, over which the needle, if free to move,
will set itself exactly vertical.   Move it back to the
centre, it resumes its horizontality ; pass it on towards
the south pole, its north end now dips, and directly
over the south pole the needle becomes vertical, its
north end being now turned downwards.   Thus we
learn that on the one side of the magnetic equator the
north end of the needle dips; on the other side the
south end dips, the dip varying from nothing to 90°.
If we go to the equatorial regions of the earth with a
suitably suspended needle we shall find there the
position of the needle horizontal.   If we sail north one
end of the needle dips; if we sail south the opposite
end dips; and over the north or south terrestrial

magnetic pole the needle sets vertical. The south magnetic pole has not yet been found, but Sir James Ross discovered the north magnetic pole on June 1, 1831. In this manner we establish a complete parallelism between the action of the earth and that of an ordinary magnet.

The terrestrial magnetic poles do not coincide with the geographical ones; nor does the earth's magnetic equator quite coincide with the geographical equator. The direction of the magnetic needle in London, which is called the magnetic meridian, encloses an angle of 24° with the astronomical meridian, this angle being called the Declination of the needle for London. The north pole of the needle now lies to the west of the true meridian; the declination is westerly. In the year 1660, however, the declination was nothing, while before that time it was easterly. All this proves that the earth's magnetic constituents are gradually changing their distribution. This change is very slow : it is therefore called the *secular change*, and the observation of it has not yet extended over a sufficient period to enable us to guess, even approximately, at its laws.

Having thus discovered, to some extent, the secret of the earth's magnetic power, we can turn it to account. In the line of 'dip' I hold a poker formed of good soft iron. The earth, acting as a magnet, is at this moment constraining the two fluids of the poker to separate, making the lower end of the poker a north pole, and the upper end a south pole. Mark the experiment : When the knob is uppermost, it attracts the north end of a magnetic needle; when undermost it attracts the south end of a magnetic needle. With such a poker repeat this experiment and satisfy yourselves that the fluids shift their position according to the manner in which the poker is presented to the earth. It has already been stated that the softest iron possesses a certain amount of coercive force. The earth, at this moment, finds in this force an antagonist which opposes the decomposition of the neutral fluid, The component fluids may be figured as meeting an amount of friction, or possessing an amount of adhesion, which prevents them from gliding over the molecules of the poker. Can we assist the earth in this case? If we wish to remove the residue of a powder from the interior surface of a glass to which the powder clings, we invert the glass, tap it, loosen the hold of the powder, and thus enable the force of gravity to pull it down. So also by tapping the end of the poker we loosen the adhesion of the magnetic fluids to the molecules and enable the earth to pull them apart. But, what is the consequence? The portion of fluid which has been thus forcibly dragged over the molecules refuses to return when the poker has been removed from the line of dip; the iron, as you see, has become a permanent magnet. By reversing its position and tapping it again we reverse its magnetism. A thoughtful and competent teacher will know how to place these remarkable facts before his pupils in a manner which will excite their interest. By the use of sensible images, more or less gross, he will first give those whom he teaches definite conceptions, purifying these conceptions afterwards, as the minds of his pupils become more capable of abstraction. By thus giving them a distinct substratum for their reasonings, he will confer upon his pupils a profit and a joy which the mere exhibition of facts without principles, or the appeal to the bodily censes and the power of memory alone, could never inspire.

As an expansion of the note at p. 357, the following extract may find a place here :—

'It is well known that a voltaic current exerts an attractive force upon a second current, flowing in the same direction ; and that when the directions are opposed to each other the force exerted is a repulsive one. By coiling wires into spirals, Ampère was enabled to make them produce all the phenomena of attraction and repulsion exhibited by magnets, and from this it was but a step to his celebrated theory of molecular currents. He supposed the molecules of a magnetic body to be surrounded by such currents, which, however, in the natural state of the body mutually neutralised each other, on account of their confused grouping. The act of magnetisation he supposed to consist in setting these molecular currents parallel to each other ; and, starting from this principle, he reduced all the phenomena of magnetism to the mutual action of electric currents.

'If we reflect upon the experiments recorded in the foregoing pages from first to last, we can hardly fail to be convinced that diamagnetic bodies operated on by magnetic forces possess a polarity "the same in kind as, but the reverse in direction of, that acquired by magnetic bodies." But if this be the case, how are we to conceive the *physical mechanism* of this polarity ? According to Coulomb's and Poisson's theory, the act of magnetisation consists in the decomposition of a neutral magnetic fluid ; the north pole of a magnet, for example, possesses an attraction for the south fluid of a piece of soft iron submitted to its influence, draws the said fluid towards it, and with it the material particles with which the fluid is associated. To account for diamagnetic phenomena this theory seems to fail altogether ; according to it, indeed, the oft-used phrase, " a north pole exciting a north pole, and a south pole a south pole," involves a contradiction. For if the north fluid be supposed to be *attracted* towards the influencing north pole, it is absurd to suppose that its presence there could produce *repulsion*. The theory of Ampère is equally at a loss to explain diamagnetic action ; for if we suppose the particles of bismuth surrounded by molecular currents, then, according to all that is known of electro-dynamic laws, these currents would set themselves parallel to, and in the same direction as, those of the magnet, and hence attraction, and not repulsion, would be the result. The fact, however, of this not being the case, proves that these molecular currents are not the mechanism by which diamagnetic induction is effected. The consciousness of this, I doubt not, drove M. Weber to the assumption that the phenomena of diamagnetism are produced by molecular currents, not *directed*, but actually *excited* in the bismuth by the magnet. Such induced currents would, according to known laws, have a direction opposed to those of the inducing magnet, and hence would produce the phenomena of repulsion. To carry out the assumption here made, M. Weber is obliged to suppose that the molecules of diamagnetic bodies are surrounded by channels, in which the induced molecular currents, once excited, continue to flow without resistance.'[1]—*Diamagnetism and Magne-crystallic Action*, p. 136-7.

---

[1] In assuming these non-resisting channels M. Weber, it must be admitted, did not go beyond the assumptions of Ampère.

XVI.

ON FORCE.[1]

A SPHERE of lead was suspended at a height of 16 feet above the theatre floor of the Royal Institution. It was liberated, and fell by gravity. That weight required a second to fall to the floor from that elevation; and the instant before it touched the floor, it had a velocity of 32 feet a second. That is to say, if at that instant the earth were annihilated, and its attraction annulled, the weight would proceed through space at the uniform velocity of 32 feet a second.

If instead of being pulled downward by gravity, the weight be cast upward in opposition to gravity, then, to reach a height of 16 feet it must start with a velocity of 32 feet a second. This velocity imparted to the weight by the human hand, or by any other mechanical means, would carry it to the precise height from which we saw it fall.

Now the lifting of the weight may be regarded as so much mechanical work performed. By means of a ladder placed against the wall, the weight might be carried up to a height of 16 feet; or it might be drawn up to this height by means of a string and pulley, or it might be suddenly jerked up to a height of 16 feet. The amount of work done in all these cases, as far as the raising of the weight is concerned, would be abso-

[1] A discourse delivered in the Royal Institution, June 6, 1862.

lutely the same. The work done at one and the same place, and neglecting the small change of gravity with the height, depends solely upon two things; on the quantity of matter lifted, and on the height to which it is lifted. If we call the quantity or mass of matter m, and the height through which it is lifted h, then the product of m into h, or $m\,h$, expresses, or is proportional to, the amount of work done.

Supposing, instead of imparting a velocity of 32 feet a second we impart at starting twice this velocity. To what height will the weight rise? You might be disposed to answer, 'To twice the height;' but this would be quite incorrect. Instead of twice 16, or 32 feet, it would reach a height of four times 16, or 64 feet. So also, if we treble the starting velocity, the weight would reach nine times the height; if we quadruple the speed at starting, we attain sixteen times the height. Thus, with a four-fold velocity of 128 feet a second at starting, the weight would attain an elevation of 256 feet. With a seven-fold velocity at starting, the weight would rise to 49 times the height, or to an elevation of 784 feet.

Now the work done—or, as it is sometimes called, the *mechanical effect*—other things being constant, is, as before explained, proportional to the height, and as a double velocity gives four times the height, a treble velocity nine times the height, and so on, it is perfectly plain that the mechanical effect increases as the square of the velocity. If the mass of the body be represented by the letter m, and its velocity by v, the mechanical effect would be proportional to or represented by $m\,v^2$. In the case considered, I have supposed the weight to be cast upward, being opposed in its flight by the resistance of gravity; but the same holds true if the projectile be sent into water, mud, earth, timber, or other resisting material. If, for example, we double the velocity of a cannon-ball, we quadruple its mechanical effect. Hence the importance of augmenting the velocity of a projectile, and hence the philosophy of Sir William Armstrong in using a large charge of powder in his recent striking experiments.

The measure then of mechanical effect is the mass of the body multiplied by the square of its velocity.

Now in firing a ball against a target the projectile, after collision, is often found hot. Mr. Fairbairn informs me that in the experiments at Shoeburyness it is a common thing to see a flash, even in broad daylight, when the ball strikes the target. And if our lead weight be examined after it has fallen from a height it is also found heated. Now here experiment and reasoning lead us to the remarkable law that, like the mechanical effect, the amount of heat generated is proportional to the product of the mass into the square of the velocity. Double your mass, other things being equal, and you double your amount of heat; double your velocity, other things remaining equal, and you quadruple your amount of heat. Here then we have common mechanical motion destroyed and heat produced. When a violin bow is drawn across a string, the sound produced is due to motion imparted to the air, and to produce that motion muscular force has been expended. We may here correctly say, that the mechanical force of the arm is converted into music. In a similar way we say that the arrested motion of our descending weight, or of the cannon-ball, is converted into heat. The mode of motion changes, but motion still continues; the motion of the mass is converted into a motion of the atoms of the mass; and these small motions, communicated to the nerves, produce the sensation we call heat.

We know the amount of heat which a given amount of mechanical force can develope. Our lead ball, for example, in falling to the earth generated a quantity of heat sufficient to raise its own temperature three-fifths of a Fahrenheit degree. It reached the earth with a velocity of 32 feet a second, and forty times this velocity would be small for a rifle bullet; multiplying ⅗ths by the square of 40, we find that the amount of heat developed by collision with the target would, if wholly concentrated in the lead, raise its temperature 960 degrees. This would be more than sufficient to fuse the lead. In reality, however, the heat developed is divided between the lead and the body against which it strikes; nevertheless, it would be worth while to pay attention to this point, and to ascertain whether rifle bullets do not, under some circumstances, show signs of fusion.[1]

From the motion of sensible masses, by gravity and other means, we now pass to the motion of atoms towards each other by chemical affinity. A collodion balloon filled with a mixture of chlorine and hydrogen being hung in the focus of a parabolic mirror, in the focus of a second mirror 20 feet distant a strong electric light was suddenly generated; the instant the concentrated light fell upon the balloon, the gases within it exploded, hydrochloric acid being the result. Here the atoms virtually fell together, the amount of heat produced showing the enormous force of the collision. The burning of charcoal in oxygen is an old experiment, but it has now a significance beyond what it used to have; we now regard the act of combination on the part of the atoms of oxygen and coal as we re-

[1] Eight years subsequently this surmise was proved correct. In the Franco-German War signs of fusion were observed in the case of bullets impinging on bones.

gard the clashing of a falling weight against the earth. The heat produced in both cases is referable to a common cause. A diamond, which burns in oxygen as a star of white light, glows and burns in consequence of the falling of the atoms of oxygen against it. And could we measure the velocity of the atoms when they clash, and could we find their number and weights, multiplying the weight of each atom by the square of its velocity, and adding all together, we should get a number representing the exact amount of heat developed by the union of the oxygen and carbon.

Thus far we have regarded the heat developed by the clashing of sensible masses and of atoms. Work is expended in giving motion to these atoms or masses, and heat is developed. But we reverse this process daily, and by the expenditure of heat execute work. We can raise a weight by heat; and in this agent we possess an enormous store of mechanical power. A pound of coal produces by its combination with oxygen an amount of heat which, if mechanically applied, would suffice to raise a weight of 100 lbs. to a height of 20 miles above the earth's surface. Conversely, 100 lbs. falling from a height of 20 miles, and striking against the earth, would generate an amount of heat equal to that developed by the combustion of a pound of coal. Wherever work is done by heat, heat disappears. A gun which fires a ball is less heated than one which fires blank cartridge. The quantity of heat communicated to the boiler of a working steam-engine is greater than that which could be obtained from the re-condensation of the steam, after it had done its work; and the amount of work performed is the exact equivalent of the amount of heat lost. Mr. Smyth informed us in his interesting discourse, that we dig annually 84 millions of tons of coal from our pits. The amount of mechanical force represented by this quantity of coal seems perfectly fabulous. The combustion of a single pound of coal, supposing it to take place in a minute, would be equivalent to the work of 300 horses; and if we suppose 108 millions of horses working day and night with unimpaired strength, for a year, their united energies would enable them to perform an amount of work just equivalent to that which the annual produce of our coal-fields would be able to accomplish.

Comparing with ordinary gravity the force with which oxygen and carbon unite together, chemical affinity seems almost infinite. But let us give gravity fair play by permitting it to act throughout its entire range. Place a body at such a distance from the earth that the attraction of our planet is barely sensible, and let it fall to the earth from this distance. It would reach the earth with a final velocity of 36,747 feet a second; and on collision with the earth the body would generate about twice the amount of heat generated by the combustion of an equal weight of coal. We have stated that by falling through a space of 16 feet our lead bullet would be heated three-fifths of a degree; but a body falling from an infinite distance has already used up 1,299,999 parts out of 1,300,000 of the earth's pulling power, when it has arrived within 16 feet of the surface; on this space only $\frac{1}{1,300,000}$ths of the whole force is exerted.

Let us now turn our thoughts for a moment from the earth to the sun. The researches of Sir John Herschel and M. Pouillet have informed us of the annual expenditure of the sun as regards heat; and by an easy calculation we ascertain the precise amount of the expenditure which falls to the share of our planet. Out of 2300 million parts of light and heat the earth receives one. The whole heat emitted by the sun in a minute would be competent to boil 12,000 millions of cubic miles of ice-cold water. How is this enormous loss made good—whence is the sun's heat derived, and by what means is it maintained? No combustion—no chemical affinity with which we are acquainted, would be competent to produce the temperature of the sun's surface. Besides, were the sun a burning body merely, its light and heat would speedily come to an end. Supposing it to be a solid globe of coal, its combustion would only cover 4600 years of expenditure. In this short time it would burn itself out. What agency then can produce the temperature and maintain the outlay? We have already regarded the case of a body falling from a great distance towards the earth, and found that the heat generated by its collision would be twice that produced by the combustion of an equal weight of coal. How much greater must be the heat developed by a body falling against the sun! .The maximum velocity with which a body can strike the earth is about 7 miles in a second; the maximum velocity with which it can strike the sun is 390 miles in a second. And as the heat developed by the collision is proportional to the square of the velocity destroyed, an asteroid falling into the sun with the above velocity would generate about 10,000 times the quantity of heat produced by the combustion of an asteroid of coal of the same weight.

Have we any reason to believe that such bodies exist in space, and that they may be raining down upon the sun? The meteorites flashing through the air are small planetary bodies, drawn by the earth's attraction. They enter our atmosphere with planetary velocity, and by friction against the air they are raised to incandescence and caused to emit light and heat. At certain

25

seasons of the year they shower down upon us in great numbers. In Boston 240,000 of them were observed in nine hours. There is no reason to suppose that the planetary system is limited to ' vast masses of enormous weight;' there is, on the contrary, reason to believe that space is stocked with smaller masses, which obey the same laws as the larger ones. That lenticular envelope which surrounds the sun, and which is known to astronomers as the Zodiacal light, is probably a crowd of meteors; and moving as they do in a resisting medium, they must continually approach the sun. Falling into it, they would produce enormous heat, and this would constitute a source from which the annual loss of heat might be made good. The sun, according to this hypothesis, would continually grow larger ; but how much larger? Were our moon to fall into the sun, it would develope an amount of heat sufficient to cover one or two years' loss; and were our earth to fall into the sun a century's loss would be made good. Still, our moon and our earth, if distributed over the surface of the sun, would utterly vanish from perception. Indeed, the quantity of matter competent to produce the required effect would, during the range of history, cause no appreciable augmentation in the sun's magnitude. The augmentation of the sun's attractive force would be more sensible. However this hypothesis may fare as a representant of what is going on in nature, it certainly shows how a sun *might* be formed and maintained on known thermo-dynamic principles.

Our earth moves in its orbit with a velocity of 68,040 miles an hour. Were this motion stopped, an amount of heat would be developed sufficient to raise the temperature of a globe of lead of the same size as the earth 384,000 degrees of the centigrade thermometer. It has been prophesied that 'the elements shall melt with fervent heat.' The earth's own motion embraces the conditions of fulfilment; stop that motion, and the greater part, if not the whole, of our planet would be reduced to vapour. If the earth fell into the sun, the amount of heat developed by the shock would be equal to that developed by the combustion of a mass of solid coal 6435 times the earth in size.

There is one other consideration connected with the permanence of our present terrestrial conditions, which is well worthy of our attention. Standing upon one of the London bridges, we observe the current of the Thames reversed, and the water poured upward twice a-day. The water thus moved rubs against the river's bed, and heat is the consequence of this friction. The heat thus generated is in part radiated into space and lost, as far as the earth is concerned. What supplies this incessant loss? The earth's rotation. Let us look a little more closely at the matter. Imagine the moon fixed, and the earth turning like a wheel from west to east in its diurnal rotation. Suppose a high mountain on the earth's surface approaching the earth's meridian; that mountain is, as it were, laid hold of by the moon; it forms a kind of handle by which the earth is pulled more quickly round. But when the meridian is passed the pull of the moon on the mountain would be in the opposite direction, it would tend to diminish the velocity of rotation as much as it previously augmented it; thus the action of all fixed bodies on the earth's surface is neutralised. But suppose the mountain to lie always to the east of the moon's meridian, the pull then would be always exerted against the earth's rotation, the velocity of which would be diminished in a degree corresponding to the strength of the pull. *The tidal wave occupies this position*—it lies always to the east of the moon's meridian. The waters of the ocean are in part dragged as a brake along the surface of the earth; and as a brake they must diminish the velocity of the earth's rotation.[1] Supposing then that we turn a mill by the action of the tide, and produce heat by the friction of the millstones; that heat has an origin totally different from the heat produced by another mill which is turned by a mountain stream. The former is produced at the expense of the earth's rotation, the latter at the expense of the sun's radiation.

The sun, by the act of vaporisation, lifts mechanically all the moisture of our air, which when it condenses falls in the form of rain, and when it freezes falls as snow. In this solid form it is piled upon the Alpine heights, and furnishes materials for glaciers. But the sun again interposes, liberates the solidified liquid, and permits it to roll by gravity to the sea. The mechanical force of every river in the world as it rolls towards the ocean, is drawn from the heat of the sun. No streamlet glides to a lower level without having been first lifted to the elevation from which it springs by the power of the sun. The energy of winds is also due entirely to the same power.

But there is still another work which the sun performs, and its connection with which is not so obvious. Trees and vegetables grow upon the earth, and when burned they give rise to heat, and hence to mechanical energy. Whence is this power derived? You see this oxide of iron, produced by the falling together of the atoms of iron and oxygen; you cannot see this transparent carbonic acid gas, formed by the falling together

---

[1] Kant surmised an action of this kind.

of carbon and oxygen. The atoms thus in close union resemble our lead weight while resting on the earth; but we can wind up the weight and prepare it for another fall, and so these atoms can be wound up and thus enabled to repeat the process of combination. In the building of plants carbonic acid is the material from which the carbon of the plant is derived; and the solar beam is the agent which tears the atoms asunder, setting the oxygen free, and allowing the carbon to aggregate in woody fibre. Let the solar rays fall upon a surface of sand; the sand is heated, and finally radiates away as much heat as it receives; let the same beams fall upon a forest, the quantity of heat given back is less than the forest receives; for the energy of a portion of the sunbeams is invested in building the trees. Without the sun the reduction of the carbonic acid cannot be effected, and an amount of sunlight is consumed exactly equivalent to the molecular work done. Thus trees are formed; thus the cotton on which Mr. Bazley discoursed last Friday is produced. I ignite this cotton, and it flames; the oxygen again unites with the carbon; but an amount of heat equal to that produced by its combustion was sacrificed by the sun to form that bit of cotton.

We cannot, however, stop at vegetable life, for it is the source, mediate or immediate, of all animal life. The sun severs the carbon from its oxygen and builds the vegetable; the animal consumes the vegetable thus formed, a reunion of the severed elements takes place, producing animal heat. The process of building a vegetable is one of winding up; the process of building an animal is one of running down. The warmth of our bodies, and every mechanical energy which we exert, trace their lineage directly to the sun.

The fight of a pair of pugilists, the motion of an
army, or the lifting of his own body by an Alpine
climber up a mountain slope, are all cases of mechanical energy drawn from the sun. A man weighing 150
pounds has 64 pounds of muscle; but these, when dried,
reduce themselves to 15 pounds. Doing an ordinary
day's work, for eighty days, this mass of muscle would
be wholly oxidised. Special organs which do more work
would be more quickly consumed: the heart, for example, if entirely unsustained, would be oxidised in
about a week. Take the amount of heat due to the
direct oxidation of a given weight of food; less heat
is developed by the oxidation of the same amount of
food in the working animal frame, and the missing
quantity is the equivalent of the mechanical work accomplished by the muscles.

I might extend these considerations; the work,
indeed, is done to my hand—but I am warned that
you have been already kept too long. To whom then
are we indebted for the most striking generalisations
of this evening's discourse? They are the work of
a man of whom you have scarcely ever heard—
the published labours of a German doctor, named
Mayer. Without external stimulus, and pursuing his
profession as town physician in Heilbronn, this man
was the first to raise the conception of the interaction
of heat and other natural forces to clearness in his
own mind. And yet he is scarcely ever heard of, and
even to scientific men his merits are but partially
known. Led by his own beautiful researches, and quite
independent of Mayer, Mr. Joule published in 1843
his first paper on the 'Mechanical Value of Heat;'
but in 1842 Mayer had actually calculated the
mechanical equivalent of heat from data which only
a man of the rarest penetration could turn to account.

In 1845 he published his memoir on 'Organic Motion,' and applied the mechanical theory of heat in the most fearless and precise manner to vital processes. He also embraced the other natural agents in his chain of conservation. In 1853 Mr. Waterston proposed, independently, the meteoric theory of the sun's heat, and in 1854 Professor William Thomson applied his admirable mathematical powers to the development of the theory; but six years previously the subject had been handled in a masterly manner by Mayer, and all that I have said about it has been derived from him. When we consider the circumstances of Mayer's life, and the period at which he wrote, we cannot fail to be struck with astonishment at what he has accomplished. Here was a man of genius working in silence, animated solely by a love of his subject, and arriving at the most important results in advance of those whose lives were entirely devoted to Natural Philosophy. It was the accident of bleeding a feverish patient at Java in 1840 that led Mayer to speculate on these subjects. He noticed that the venous blood in the tropics was of a brighter red than in colder latitudes, and his reasoning on this fact led him into the laboratory of natural forces, where he has worked with such signal ability and success. Well, you will desire to know what has become of this man. His mind, it is alleged, gave way; it is said he became insane, and he was certainly sent to a lunatic asylum. In a biographical dictionary of his country it is stated that he died there, but this is incorrect. He recovered; and, I believe, is at this moment a cultivator of vineyards in Heilbronn.

June 20, 1862.

While preparing for publication my last course of lectures on Heat, I wished to make myself acquainted with all that Dr. Mayer had done in connection with this subject. I accordingly wrote to two gentlemen who above all others seemed likely to give me the information which I needed.[1]   Both of them are Germans, and both particularly distinguished in connection with the Dynamical Theory of Heat.   Each of them kindly furnished me with the list of Mayer's publications, and one of them [Clausius] was so friendly as to order them from a bookseller, and to send them to me.   This friend, in his reply to my first letter regarding Mayer, stated his belief that I should not find anything very important in Mayer's writings; but before forwarding the memoirs to me he read them himself.   His letter accompanying them contains the following words:—
'I must here retract the statement in my last letter, that you would not find much matter of importance in Mayer's writings: I am astonished at the multitude of beautiful and correct thoughts which they contain;' and he goes on to point out various important subjects, in the treatment of which Mayer had anticipated other eminent writers.   My other friend, in whose own publications the name of Mayer repeatedly occurs, and whose papers containing these references were translated some years ago by myself, was, on the 10th of last month, unacquainted with the thoughtful and beautiful essay of Mayer's, entitled 'Beiträge zur Dynamik des Himmels,' and in 1854, when Professor William Thomson developed in so striking a manner the meteoric theory of the sun's heat, he was certainly not aware of the existence of that essay, though from a recent article in

[1] Helmholtz and Clausius.

'Macmillan's Magazine' I infer that he is now aware of it. Mayer's physiological writings have been referred to by physiologists—by Dr. Carpenter, for example —in terms of honouring recognition. We have hitherto, indeed, obtained fragmentary glimpses of the man, partly from physicists and partly from physiologists; but his total merit has never yet been recognised as it assuredly would have been had he chosen a happier mode of publication. I do not think a greater disservice could be done to a man of science, than to overstate his claims: such overstatement is sure to recoil to the disadvantage of him in whose interest it is made. But when Mayer's opportunities, achievements, and fate are taken into account, I do not think that I shall be deeply blamed for attempting to place him in that honourable position, which I believe to be his due.

Here, however, are the titles of Mayer's papers, the perusal of which will correct any error of judgment into which I may have fallen regarding their author. 'Bemerkungen über die Kräfte der unbelebten Natur,' Liebig's 'Annalen,' 1842, Vol. 42, p. 231 ; 'Die Organische Bewegung in ihrem Zusammenhange mit dem Stoffwechsel,' Heilbronn, 1845; 'Beiträge zur Dynamik des Himmels,' Heilbronn, 1848 ; 'Bemerkungen über das Mechanische Equivalent der Wärme,' Heilbronn, 1851.

IN MEMORIAM.—Dr. Julius Robert Mayer died at Heilbronn on March 20, 1878, aged 63 years. It gives me pleasure to reflect that the great position which he will for ever occupy in the annals of science was first virtually assigned to him in the foregoing discourse. He was subsequently chosen by acclamation a member of the French Academy of Sciences ; and he received from

the Royal Society the Copley medal—its highest
reward.[1]

November 1878.

At the meeting of the British Association at Glas-
gow in 1876—that is to say, more than fourteen years
after its delivery and publication—the foregoing lecture
was made the cloak for an unseemly personal attack by
Professor Tait.   The anger which found this uncour-
teous vent dates from 1863,[2] when it fell to my lot to
maintain, in opposition to him and a more eminent
colleague, the position which in 1862 I had assigned to
Dr. Mayer.   In those days Professor Tait denied to
Mayer all originality, and he has since, I regret to say,
never missed an opportunity, however small, of carping
at Mayer's claims.   The action of the Academy of
Sciences and of the Royal Society summarily disposes
of this detraction, to which its object, during his life-
time, never vouchsafed either remonstrance or reply.

    Some time ago Professor Tait published a volume of
lectures entitled ' Recent Advances in Physical Science,'
which I have reason to know has evoked an amount of
censure far beyond that hitherto publicly expressed.
Many of the best heads on the continent of Europe
agree in their rejection and condemnation of the historic
portions of this book.   In March last it was subjected
to a brief but pungent critique by Du Bois-Reymond,
the celebrated Perpetual Secretary of the Academy of
Sciences in Berlin.   Du Bois-Reymond's address was
on ' National Feeling,' and his critique is thus wound
up :—' The author of the " Lectures " is not, perhaps,

---

[1] See ' The Copley Medalist for 1871,' p. 479.
[2] See ' Philosophical Magazine ' for this and the succeeding years.

sufficiently well acquainted with the history on which he professes to throw light, and on the later phases of which he passes so unreserved (schroff) a judgment. He thus exposes himself to the suspicion—which, unhappily, is not weakened by his other writings—that the fiery Celtic blood of his country occasionally runs away with him, converting him for the time into a scientific Chauvin. Scientific Chauvinism,' adds the learned secretary, 'from which German investigators have hitherto kept free, is more reprehensible (gehässig) than political Chauvinism, inasmuch as self-control (*sittliche Haltung*) is more to be expected from men of science, than from the politically excited mass.'

In the case before this 'expectation' would, I fear, be doomed to disappointment. But Du Bois-Reymond and his countrymen must not accept the writings of Professor Tait as representative of the thought of England. Surely no nation in the world has more effectually shaken itself free from scientific Chauvinism. From the day that Davy, on presenting the Copley medal to Arago, scornfully brushed aside that spurious patriotism which would run national boundaries through the free domain of science, chivalry towards foreigners has been a guiding principle with the Royal Society.

On the more private amenities indulged in by Professor Tait, I do not consider it necessary to say a word.

---

[1] Festrede, delivered before the Academy of Sciences of Berlin, in celebration of the birthday of the Emperor and King, March 28, 1878.

XVII.

CONTRIBUTIONS TO MOLECULAR PHYSICS.[1]

HAVING on previous occasions dwelt upon the enormous differences which exist among gaseous bodies both as regards their power of absorbing and emitting radiant heat, I have now to consider the effect of a change of aggregation. When a gas is condensed to a liquid, or a liquid congealed to a solid, the molecules coalesce, and grapple with each other by forces which are insensible as long as the gaseous state is maintained. But, even in the solid and liquid conditions, the luminiferous ether still surrounds the molecules: hence, if the acts of radiation and absorption depend on them individually, regardless of their state of aggregation, the change from the gaseous to the liquid state ought not materially to affect the radiant and absorbent power. If, on the contrary, the mutual entanglement of the molecules by the force of cohesion be of paramount influence, then we may expect that liquids will exhibit a deportment towards radiant heat altogether different from that of the vapours from which they are derived.

The first part of an enquiry conducted in 1863-64 was devoted to an exhaustive examination of this question. Twelve different liquids were employed, and five

[1] A discourse delivered at the Royal Institution, March 18, 1864— supplementing, though of prior date, the Rede Lecture on Radiation.

different layers of each, varying in thickness from 0·02 of an inch to 0·27 of an inch.   The liquids were enclosed, not in glass vessels, which would have materially modified the incident heat, but between plates of transparent rock-salt, which only slightly affected the radiation. The source of heat throughout these comparative experiments consisted of a platinum wire, raised to incandescence by an electric current of unvarying strength.   The quantities of radiant heat absorbed and transmitted by each of the liquids at the respective thicknesses were first determined.   The vapours of these liquids were subsequently examined, the quantities of vapour employed being rendered proportional to the quantities of liquid previously traversed by the radiant heat.   The result was that, for heat from the same source, the order of absorption of liquids and of their vapours proved absolutely the same.   There is no known exception to this law; so that, to determine the position of a vapour as an absorber or a radiator, it is only necessary to determine the position of its liquid.

This result proves that the state of aggregation, as far at all events as the liquid stage is concerned, is of altogether subordinate moment—a conclusion which will probably prove to be of cardinal importance in molecular physics.   On one important and contested point it has a special bearing.   If the position of a liquid as an absorber and radiator determine that of its vapour, the position of water fixes that of aqueous vapour. Water has been compared with other liquids in a multitude of experiments, and it has been found, both as a radiant and as an absorbent, to transcend them all. Thus, for example, a layer of bisulphide of carbon 0·02 of an inch in thickness absorbs 6 per cent., and allows 94 per cent. of the radiation from the red-hot platinum spiral to pass through it; benzol absorbs 43 and transmits 57 per cent. of the same radiation; alcohol absorbs 67 and transmits 33 per cent., and alcohol, as an absorber of radiant heat, stands at the head of all liquids except one. The exception is water. A layer of this substance, of the thickness above given, absorbs 81 per cent., and permits only 19 per cent. of the radiation to pass through it. Had no single experiment ever been made upon the vapour of water, its vigorous action upon radiant heat might be inferred from the deportment of the liquid.

The relation of absorption and radiation to the chemical constitution of the radiating and absorbing substances was next briefly considered. For the first six substances in the list of liquids examined, the radiant and absorbent powers augment as the number of atoms in the compound molecule augments. Thus, bisulphide of carbon has 3 atoms, chloroform 5, iodide of ethyl 8, benzol 12, and amylene 15 atoms in their respective molecules. The order of their power as radiants and absorbents is that here indicated, bisulphide of carbon being the feeblest, and amylene the strongest of the six. Alcohol, however, excels benzol as an absorber, though it has but 9 atoms in its molecule; but, on the other hand, its molecule is rendered more complex by the introduction of a new element. Benzol contains carbon and hydrogen, while alcohol contains carbon, hydrogen and oxygen. Thus, not only does atomic *multitude* come into play in absorption and radiation —atomic *complexity* must also be taken into account. I would recommend to the particular attention of chemists the molecule of water; the deportment of this substance towards radiant heat being perfectly anomalous, if the chemical formula at present ascribed to it be correct.

Sir William Herschel made the important discovery

that, beyond the limits of the red end of the solar spectrum, rays of high heating power exist which are incompetent to excite vision. The discovery is capable of extension. Dissolving iodine in the bisulphide of carbon, a solution is obtained which entirely intercepts the light of the most brilliant flames, while to the ultra-red rays of such flames the same iodine is found to be perfectly diathermic. The transparent bisulphide, which is highly pervious to invisible heat, exercises on it the same absorption as the perfectly opaque solution. A hollow prism filled with the opaque liquid being placed in the path of the beam from an electric lamp, the light-spectrum is completely intercepted, but the heat-spectrum may be received upon a screen and there examined. Falling upon a thermo-electric pile, its invisible presence is shown by the prompt deflection of even a coarse galvanometer.

What, then, is the physical meaning of opacity and transparency as regards light and radiant heat? The visible rays of the spectrum differ from the invisible ones simply in *period*. The sensation of light is excited by waves of ether shorter and more quickly recurrent than the non-visual waves which fall beyond the extreme red. But why should iodine stop the former and allow the latter to pass? The answer to this question no doubt is, that the intercepted waves are those whose periods of recurrence coincide with the periods of oscillation possible to the atoms of the dissolved iodine. The elastic forces which keep these atoms apart compel them to vibrate in definite periods, and, when these periods synchronise with those of the ethereal waves, the latter are absorbed. Briefly defined, then, transparency in liquids, as well as in gases, is synonymous with *discord*, while opacity is synonymous with *accord*, between the periods of the waves of

ether and those of the molecules on which they impinge.

According to this view transparent and colourless
substances owe their transparency to the dissonance
existing between the oscillating periods of their atoms
and those of the waves of the whole visible spectrum. From the prevalence of transparency in compound bodies, the general discord of the vibrating
periods of their atoms with the light-giving waves
of the spectrum, may be inferred; while their synchronism with the ultra-red periods is to be inferred from
their opacity to the ultra-red rays. Water illustrates
this in a most striking manner. It is highly transparent to the luminous rays, which proves that its
atoms do not readily oscillate in the periods which
excite vision. It is highly opaque to the ultra-red
undulations, which proves the synchronism of its vibrating periods with those of the longer waves.

If, then, to the radiation from any source water
shows itself eminently or perfectly opaque, we may
infer that the atoms whence the radiation emanates
oscillate in ultra-red periods. Let us apply this test
to the radiation from a flame of hydrogen. This
flame consists mainly of incandescent aqueous vapour,
the temperature of which, as calculated by Bunsen,
is 3259° C., so that, if the penetrative power of
radiant heat, as generally supposed, augment with the
temperature of its source, we may expect the radiation from this flame to be copiously transmitted by
water. While, however, a layer of the bisulphide
of carbon 0·07 of an inch in thickness transmits 72 per
cent. of the incident radiation, and while every other
liquid examined transmits more or less of the heat, a
layer of water of the above thickness is entirely opaque to
the radiation from the hydrogen flame. Thus we establish

accord between the periods of the atoms of cold water and those of aqueous vapour at a temperature of 3259° C. But the periods of water have already been proved to be ultra red—hence those of the hydrogen flame must be sensibly ultra-red also. The absorption by dry air of the heat emitted by a platinum spiral raised to incandescence by electricity is insensible, while that by the ordinary undried air is 6 per cent. Substituting for the platinum spiral a hydrogen flame, the absorption by dry air still remains insensible, while that of the undried air rises to 20 per cent. of the entire radiation. The temperature of the hydrogen flame is, as stated, 3259° C.; that of the aqueous vapour of the air 20° C. Suppose, then, the temperature of aqueous vapour to rise from 20° C. to 3259° C., we must conclude that the augmentation of temperature is applied to an increase of amplitude or width of swing, and not to the introduction of quicker periods into the radiation.

The part played by aqueous vapour in the economy of nature is far more wonderful than has been hitherto supposed. To nourish the vegetation of the earth the actinic and luminous rays of the sun must penetrate our atmosphere; and to such rays aqueous vapour is eminently transparent. The violet and the ultra-violet rays pass through it with freedom. To protect vegetation from destructive chills the terrestrial rays must be checked in their transit towards stellar space; and this is accomplished by the aqueous vapour diffused through the air. This substance is the great moderator of the earth's temperature, bringing its extremes into proximity, and obviating contrasts between day and night which would render life insupportable. But we can advance beyond this general statement, now that we know the radiation from aqueous vapour is intercepted, in a special degree, by water, and, reciprocally, the radiation from water by aqueous vapour; for it follows from this that the very act of nocturnal refrigeration which produces the condensation of aqueous vapour at the surface of the earth—giving, as it were, a varnish of water to that surface—imparts to terrestrial radiation that particular character which disqualifies it from passing through the earth's atmosphere and losing itself in space.

And here we come to a question in molecular physics which at the present moment occupies attention. By allowing the violet and ultra-violet rays of the spectrum to fall upon sulphate of quinine and other substances Professor Stokes has changed the periods of those rays. Attempts have been made to produce a similar result at the other end of the spectrum—to convert the ultra-red periods into periods competent to excite vision—but hitherto without success. Such a change of period, I agree with Dr. Miller in believing, occurs when the lime-light is produced by an oxy-hydrogen flame. In this common experiment there is an actual breaking up of long periods into short ones—a true rendering of unvisual periods visual. The change of refrangibility here effected differs from that of Professor Stokes; firstly, by its being in the opposite direction—that is, from a lower refrangibility to a higher; and, secondly, in the circumstance that the lime is heated by the collision of the molecules of aqueous vapour, before their heat has assumed the radiant form. But it cannot be doubted that the same effect would be produced by radiant heat of the same periods, provided the motion of the æther could be rendered sufficiently intense.[1] The effect in principle is the same, whether we consider the lime to be struck by

---

[1] This was soon afterwards accomplished. See pp. 48, 49.

a particle of aqueous vapour oscillating at a certain rate, or by a particle of ether oscillating at the same rate.

By plunging a platinum wire into a hydrogen flame we cause it to glow, and thus introduce shorter periods into the radiation. These, as already stated, are in discord with the atomic vibrations of water; hence we may infer that the transmission through water will be rendered more copious by the introduction of the wire into the flame. Experiment proves this conclusion to be true. Water, from being opaque, opens a passage to 6 per cent. of the radiation from the spiral. A thin plate of colourless glass, moreover, transmits 58 per cent. of the radiation from the hydrogen flame; but when the flame and spiral are employed, 78 per cent. of the heat is transmitted.

For an alcohol flame Knoblauch and Melloni found glass to be less transparent than for the same flame with a platinum spiral immersed in it; but Melloni afterwards showed that the result was not general —that black glass and black mica were decidedly more diathermic to the radiation from the pure alcohol flame. Melloni did not explain this, but the reason is now obvious. The mica and glass owe their blackness to the carbon diffused through them. This carbon, as first proved by Melloni, is in some measure transparent to the ultra-red rays, and I have myself succeeded in transmitting between 40 and 50 per cent. of the radiation from a hydrogen flame through a layer of carbon which intercepted the light of an intensely brilliant flame. The products of combustion of alcohol are carbonic acid and aqueous vapour, the heat of which is almost wholly ultra-red. For this radiation, then, the carbon is in a considerable degree transparent, while for the radiation from the platinum spiral, it is in a great measure

opaque.   The platinum wire, therefore, which augmented the radiation through the pure glass, augmented
the absorption of the black glass and mica.

No more striking or instructive illustration of the
influence of coincidence could be adduced than that
furnished by the radiation from a carbonic oxide flame.
Here the product of combustion is carbonic acid; and
on the radiation from this flame even the ordinary
carbonic acid of the atmosphere exerts a powerful
effect.   A quantity of the gas, only one-thirtieth of an
atmosphere in density, contained in a polished brass
tube four feet long, intercepts 50 per cent. of the
radiation from the carbonic oxide flame.   For the heat
emitted by lampblack, olefiant gas is a far more
powerful absorber than carbonic acid; in fact, for such
heat, with one exception, carbonic acid is the most
feeble absorber to be found among the compound gases.
Moreover, for the radiation from a hydrogen flame
olefiant gas possesses twice the absorbent power of
carbonic acid, while for the radiation from the carbonic
oxide flame, at a common pressure of one inch of mercury, the absorption by carbonic acid is more than
twice that of olefiant gas.   Thus we establish the
coincidence of period between carbonic acid at a temperature of 20° C. and carbonic acid at a temperature
of over 3000° C., the periods of oscillation of both the
incandescent and the cold gas belonging to the ultra-
red portion of the spectrum.

It will be seen from the foregoing remarks and
experiments how impossible it is to determine the effect
of temperature pure and simple on the transmission of
radiant heat if different sources of heat be employed.
Throughout such an examination the same oscillating
atoms ought to be retained.   This is done by heating a
platinum spiral by an electric current, the temperature

meanwhile varying between the widest possible limits. Their comparative opacity to the ultra-red rays shows the general accord of the oscillating periods of the vapours referred to at the commencement of this lecture with those of the ultra-red undulations. Hence, by gradually heating a platinum wire from darkness up to whiteness, we ought gradually to augment the discord between it and these vapours, and thus augment the transmission. Experiment entirely confirms this conclusion. Formic ether, for example, absorbs 45 per cent. of the radiation from a platinum spiral heated to barely visible redness; 32 per cent. of the radiation from the same spiral at a red heat; 26 per cent. of the radiation from a white-hot spiral, and only 21 per cent. when the spiral is brought near its point of fusion. Remarkable cases of inversion as to transparency also occur. For barely visible redness formic ether is more opaque than sulphuric; for a bright red heat both are equally transparent; while, for a white heat, and still more for a higher temperature, sulphuric ether is more opaque than formic. This result gives us a clear view of the relationship of the two substances to the luminiferous ether. As we introduce waves of shorter period the sulphuric æther augments most rapidly in opacity; that is to say, its accord with the shorter waves is greater than that of the formic. Hence we may infer that the atoms of formic ether oscillate, on the whole, more slowly than those of sulphuric ether.

When the source of heat is a Leslie's cube coated with lampblack and filled with boiling water, the opacity of formic æther in comparison with sulphuric is very decided. With this source also the positions of chloroform and iodide of methyl are inverted. For a white-hot spiral, the absorption of chloroform vapour being 10 per cent., that of iodide of methyl is

16 ; with the blackened cube as source, the absorption by chloroform is 22 per cent., while that by the iodide of methyl is only 19. This inversion is not the result of temperature merely; for when a platinum wire, heated to the temperature of boiling water, is employed as a source, the iodide continues to be the most powerful absorber. All the experiments hitherto made go to prove that from heated lampblack an emission takes place which synchronises in an especial manner with chloroform. For the cube at 100° C., coated with lampblack, the absorption by chloroform is more than three times that by bisulphide of carbon ; for the radiation from the most luminous portion of a gas-flame the absorption by chloroform is also considerably in excess of that by bisulphide of carbon ; while, for the flame of a Bunsen's burner, from which the incandescent carbon particles are removed by the free admixture of air, the absorption by bisulphide of carbon is nearly twice that by chloroform. *The removal of the carbon particles more than doubles the relative transparency of the chloroform.* Testing, moreover, the radiation from various parts of the same flame, it was found that for the blue base of the flame the bisulphide of carbon was most opaque, while for all other parts of the flame the chloroform was most opaque. For the radiation from a very small gas flame, consisting of a blue base and a small white tip, the bisulphide was also most opaque, and its opacity very decidedly exceeded that of the chloroform when the source of heat was the flame of bisulphide of carbon. Comparing the radiation from a Leslie's cube coated with isinglass with that from a similar cube coated with lampblack, at the common temperature of 100° C., it was found that, out of eleven vapours, all but one absorbed the radiation from the isinglass most powerfully ; the single exception was chloroform.

It is worthy of remark that whenever, through a change of source, the position of a vapour as an absorber of radiant heat was altered, the position of the liquid from which the vapour was derived underwent a similar change.

It is still a point of difference between eminent investigators whether radiant heat, up to a temperature of 100° C., is monochromatic or not. Some affirm this; some deny it. A long series of experiments enables me to state that probably no two substances at a temperature of 100° C. emit heat of the same quality. The heat emitted by isinglass, for example, is different from that emitted by lampblack, and the heat emitted by cloth, or paper, differs from both. It is also a subject of discussion whether rock-salt is equally diathermic to all kinds of calorific rays; the differences affirmed to exist by some investigators being ascribed by others to differences of incidence from the various sources employed. MM. de la Provostaye and Desains maintain the former view, Melloni and M. Knoblauch maintain the latter. I tested this point without changing anything but the temperature of the source; its size, distance, and surroundings remaining the same. The experiments proved rock-salt to be coloured thermally. It is more opaque, for example, to the radiation from a barely visible spiral than to that from a white-hot one.

In regard to the relation of radiation to conduction, if we define radiation, internal as well as external, as the communication of motion from the vibrating atoms to the ether, we may, I think, by fair theoretic reasoning, reach the conclusion that the best radiators ought to prove the worst conductors. A broad consideration of the subject shows at once the general harmony of this conclusion with observed facts. Organic substances are all excellent radiators; they are also extremely bad conductors. The moment we pass from the metals to their compounds we pass from good conductors to bad ones, and from bad radiators to good ones. Water, among liquids, is probably the worst conductor; it is the best radiator. Silver, among solids, is the best conductor; it is the worst radiator. The excellent researches of MM. de la Provostaye and Desains furnish a striking illustration of what I am inclined to regard as a natural law—that those atoms which transfer the greatest amount of motion to the ether, or, in other words, radiate most powerfully, are the least competent to communicate motion to each other, or, in other words, to propagate by conduction readily.

XVIII.

LIFE AND LETTERS OF FARADAY.

1870.

UNDERTAKEN and executed in a reverent and loving spirit, the work of Dr. Bence Jones makes Faraday the virtual writer of his own life. Everybody now knows the story of the philosopher's birth; that his father was a smith; that he was born at Newington Butts in 1791; that he ran along the London pavements, a bright-eyed errand boy, with a load of brown curls upon his head and a packet of newspapers under his arm; that the lad's master was a bookseller and bookbinder—a kindly man, who became attached to the little fellow, and in due time made him his apprentice without fee; that during his apprenticeship he found his appetite for knowledge provoked and strengthened by the books he stitched and covered. Thus he grew in wisdom and stature to his year of legal manhood, when he appears in the volumes before us as a writer of letters, which reveal his occupation, acquirements, and tone of mind. His correspondent was Mr. Abbott, a member of the Society of Friends, who, with a forecast of his correspondent's greatness, preserved his letters and produced them at the proper time.

In later years Faraday always carried in his pocket a blank card, on which he jotted down in pencil his thoughts and memoranda. He made his notes in the laboratory, in the theatre, and in the streets. This distrust of his memory reveals itself in his first letter to Abbot. To a proposition that no new enquiry should be started between them before the old one had been exhaustively discussed, Faraday objects. 'Your notion,' he says, 'I can hardly allow, for the following reason : ideas and thoughts spring up in my mind which are irrevocably lost for want of noting at the time.' Gentle as he seemed, he wished to have his own way, and he had it throughout his life. Differences of opinion sometimes arose between the two friends, and then they resolutely faced each other. 'I accept your offer to fight it out with joy, and shall in the battle of experience cause not pain, but, I hope, pleasure.' Faraday notes his own impetuosity, and incessantly checks it. There is at times something almost mechanical in his self-restraint. In another nature it would have hardened into mere 'correctness' of conduct; but his overflowing affections prevented this in his case. The habit of self-control became a second nature to him at last, and lent serenity to his later years.

In October 1812 he was engaged by a Mr. De la Roche as a journeyman bookbinder ; but the situation did not suit him. His master appears to have been an austere and passionate man, and Faraday was to the last degree sensitive. All his life he continued so. He suffered at times from dejection; and a certain grimness, too, pervaded his moods. 'At present,' he writes to Abbott, 'I am as serious as you can be, and would not scruple to speak a truth to any human being, whatever repugnance it might give rise to. Being in this state of mind, I should have refrained from writing to you, did I not conceive from the general tenor of your letters that your mind is, at proper times, occupied upon

serious subjects to the exclusion of those that are
frivolous.' Plainly he had fallen into that stern Puritan
mood, which not only crucifies the affections and lusts
of him who harbours it, but is often a cause of disturbed
digestion to his friends.

About three months after his engagement with
De la Roche, Faraday quitted him and bookbinding
together. He had heard Davy, copied his lectures, and
written to him, entreating to be released from Trade,
which he hated, and enabled to pursue Science. Davy
recognised the merit of his correspondent, kept his eye
upon him, and, when occasion offered, drove to his door
and sent in a letter, offering him the post of assistant
in the laboratory of the Royal Institution. He was
engaged March 1, 1813, and on the 8th we find him
extracting the sugar from beet-root. He joined the
City Philosophical Society which had been founded by
Mr. Tatum in 1808. 'The discipline was very sturdy,
the remarks very plain, and the results most valuable.'
Faraday derived great profit from this little association.
In the laboratory he had a discipline sturdier still.
Both Davy and himself were at this time frequently cut
and bruised by explosions of chloride of nitrogen. One
explosion was so rapid 'as to blow my hand open, tear
away a part of one nail, and make my fingers so sore
that I cannot use them easily.' In another experiment
'the tube and receiver were blown to pieces, I got a cut
on the head, and Sir Humphry a bruise on his hand.'
And again speaking of the same substance, he says,
'when put in the pump and exhausted, it stood for a
moment, and then exploded with a fearful noise. Both
Sir H. and I had masks on, but I escaped this time the
best. Sir H. had his face cut in two places about the
chin, and a violent blow on the forehead struck through

a considerable thickness of silk and leather.' It was this same substance that blew out the eye of Dulong.

Over and over again, even at this early date, we can discern the quality which, compounded with his rare intellectual power, made Faraday a great experimental philosopher. This was his desire to see facts, and not to rest contented with the descriptions of them. He frequently pits the eye against the ear, and affirms the enormous superiority of the organ of vision. Late in life I have heard him say that he could never fully understand an experiment until he had seen it. But he did not confine himself to experiment. He aspired to be a teacher, and reflected and wrote upon the method of scientific exposition. 'A lecturer,' he observes, ' should appear easy and collected, undaunted and unconcerned :' still ' his whole behaviour should evince respect for his audience.' These recommendations were afterwards in great part embodied by himself. I doubt his 'unconcern,' but his fearlessness was often manifested. It used to rise within him as a wave, which carried both him and his audience along with it. On rare occasions also, when he felt himself and his subject hopelessly unintelligible, he suddenly evoked a certain recklessness of thought, and, without halting to extricate his bewildered followers, he would dash alone through the jungle into which he had unwittingly led them ; thus saving them from ennui by the exhibition of a vigour which, for the time being, they could neither share nor comprehend.

In October 1813 he quitted England with Sir Humphry and Lady Davy. During his absence he kept a journal, from which copious and interesting extracts have been made by Dr. Bence Jones. Davy was considerate, preferring at times to be his own servant rather than impose on Faraday duties which he disliked. But Lady

Davy was the reverse. She treated him as an underling ; he chafed under the treatment, and was often on the point of returning home. They halted at Geneva. De la Rive, the elder, had known Davy in 1799, and, by his writings in the ' Bibliothèque Britannique,' had been the first to make the English chemist's labours known abroad. He welcomed Davy to his country residence in 1814. Both were sportsmen, and they often went out shooting together. On these occasions Faraday charged Davy's gun while De la Rive charged his own. Once the Genevese philosopher found himself by the side of Faraday, and in his frank and genial way entered into conversation with the young man. It was evident that a person possessing such a charm of manner and such high intelligence could be no mere servant. On enquiry De la Rive was somewhat shocked to find that the *soidisant domestique* was really *préparateur* in the laboratory of the Royal Institution; and he immediately proposed that Faraday thenceforth should join the masters instead of the servants at their meals. To this Davy, probably out of weak deference to his wife, objected ; but an arrangement was come to that Faraday thenceforward should have his food in his own room. Rumour states that a dinner in honour of Faraday was given by De la Rive. This is a delusion ; there was no such banquet ; but Faraday never forgot the kindness of the friend who saw his merit when he was a mere *garçon de laboratoire.*[1]

[1] While confined last autumn at Geneva by the effects of a fall in the Alps, my friends, with a kindness I can never forget, did all that friendship could suggest to render my captivity pleasant to me. M. de la Rive then wrote out for me the full account, of which the foregoing is a condensed abstract. It was at the desire of Dr. Bence Jones that I asked him to do so. The rumour of a banquet at Geneva illustrates the tendency to substitute for the youth of 1814 the Faraday of later years.

He returned in 1815 to the Royal Institution. **Here**
he helped Davy for years; he worked also for himself,
and lectured frequently at the City Philosophical Society.
He took lessons in elocution, happily without damage to
his natural force, earnestness, and grace of delivery. He
was never pledged to theory, and he changed in opinion
as knowledge advanced. With him life was growth.
In those early lectures we hear him say, ' In knowledge,
that man only is to be contemned and despised who is
not in a state of transition.' And again : ' Nothing is
more difficult and requires more caution than philosophical deduction, nor is there anything more adverse
to its accuracy than fixity of opinion.' Not that he was
wafted about by every wind of doctrine; but that he
united flexibility with his strength. In striking contrast with this intellectual expansiveness was his fixity
in religion, but this is a subject which cannot be discussed here.

Of all the letters published in these volumes none
possess a greater charm than those of Faraday to his
wife. Here, as Dr. Bence Jones truly remarks, ' he laid
open all his mind and the whole of his character, and
what can be made known can scarcely fail to charm every
one by its loveliness, its truthfulness, and its earnestness.' Abbott and he sometimes swerved into word-
play about love ; but up to 1820, or thereabouts, the
passion was potential merely. Faraday's journal indeed
contains entries which show that he took pleasure in the
assertion of his contempt for love ; but these very
entries became links in his destiny. It was through
them that he became acquainted with one who inspired
him with a feeling which only ended with his life. His
biographer has given us the means of tracing the varying moods which preceded his acceptance. They reveal
more than the common alternations of light and gloom ;

at one moment he wishes that his flesh might melt and that he might become nothing; at another he is intoxicated with hope. The impetuosity of his character was then unchastened by the discipline to which it was subjected in after years. The very strength of his passion proved for a time a bar to its advance, suggesting, as it did, to the conscientious mind of Miss Barnard, doubts of her capability to return it with adequate force. But they met again and again, and at each successive meeting he found his heaven clearer, until at length he was able to say, ' Not a moment's alloy of this evening's happiness occurred. Everything was delightful to the last moment of my stay with my companion, because she was so.' The turbulence of doubt subsided, and a calm and elevating confidence took its place. ' What can I call myself,' he writes to her in a subsequent letter, ' to convey most perfectly my affection and love for you? Can I or can truth say more than that for this world I am yours?' Assuredly he made his profession good, and no fairer light falls upon his character than that which reveals his relations to his wife. Never, I believe, existed a manlier, purer, steadier love. Like a burning diamond, it continued to shed, for six-and-forty years, its white and smokeless glow.

Faraday was married on June 12, 1821; and up to this date Davy appears throughout as his friend. Soon afterwards, however, disunion occurred between them, which, while it lasted, must have given Faraday intense pain. It is impossible to doubt the honesty of conviction with which this subject has been treated by Dr. Bence Jones, and there may be facts known to him, but not appearing in these volumes, which justify his opinion that Davy in those days had become jealous of Faraday. This, which is the prevalent belief, is also reproduced in an excellent article in the March number of ' Fraser's

Magazine.' But the best analysis I can make of the data fails to present Davy in this light to me. The facts, as I regard them, are briefly these.

In 1820, Oersted of Copenhagen made the celebrated discovery which connects electricity with magnetism, and immediately afterwards the acute mind of Wollaston perceived that a wire carrying a current ought to rotate round its own axis under the influence of a magnetic pole. In 1821 he tried, but failed, to realise this result in the laboratory of the Royal Institution. Faraday was not present at the moment, but he came in immediately afterwards and heard the conversation of Wollaston and Davy about the experiment. He had also heard a rumour of a wager that Dr. Wollaston would eventually succeed.

This was in April. In the autumn of the same year Faraday wrote a history of electro-magnetism, and repeated for himself the experiments which he described. It was while thus instructing himself that he succeeded in causing a wire, carrying an electric current, to rotate round a magnetic pole. This was not the result sought by Wollaston, but it was closely related to that result.

The strong tendency of Faraday's mind to look upon the reciprocal actions of natural forces gave birth to his greatest discoveries; and we, who know this, should be justified in concluding that, even had Wollaston not preceded him, the result would have been the same. But in judging Davy we ought to transport ourselves to his time, and carefully exclude from our thoughts and feelings that noble subsequent life, which would render simply impossible the ascription to Faraday of anything unfair. It would be unjust to Davy to put our knowledge in the place of his, or to credit him with data which he could not have possessed. Rumour and fact had connected the name of Wollaston with these supposed interactions between magnets and currents. When, therefore, Faraday in October published his successful experiment, without any allusion to Wollaston, general, though really ungrounded, criticism followed. I say ungrounded because, firstly, Faraday's experiment was not that of Wollaston, and secondly, Faraday, before he published it, had actually called upon Wollaston, and not finding him at home, did not feel himself authorised to mention his name.

In December, Faraday published a second paper on the same subject, from which, through a misapprehension, the name of Wollaston was also omitted. Warburton and others thereupon affirmed that Wollaston's ideas had been appropriated without acknowledgment, and it is plain that Wollaston himself, though cautious in his utterance, was also hurt. Censure grew till it became intolerable. 'I hear,' writes Faraday to his friend Stodart, 'every day more and more of these sounds, which, though only whispers to me, are, I suspect, spoken aloud among scientific men.' He might have written explanations and defences, but he went straighter to the point. He wished to see the principals face to face—to plead his cause before them personally. There was a certain vehemence in his desire to do this. He saw Wollaston, he saw Davy, he saw Warburton; and I am inclined to think that it was the irresistible candour and truth of character which these *vivâ voce* defences revealed, as much as the defences themselves, that disarmed resentment at the time.

As regards Davy, another cause of dissension arose in 1823. In the spring of that year Faraday analysed the hydrate of chlorine, a substance once believed to be the element chlorine, but proved by Davy to be a

27

compound of that element and water. The analysis
was looked over by Davy, who then and there suggested
to Faraday to heat the hydrate in a closed glass tube.
This was done, the substance was decomposed, and one
of the products of decomposition was proved by Faraday
to be chlorine liquefied by its own pressure. On the
day of its discovery he communicated this result to
Dr. Paris. Davy, on being informed of it, instantly
liquefied another gas in the same way. Having struck
thus into Faraday's enquiry, ought he not to have left
the matter in Faraday's hands? I think he ought.
But, considering his relation to both Faraday and the
hydrate of chlorine, Davy, I submit, may be excused
for thinking differently. A father is not always wise
enough to see that his son has ceased to be a boy, and
estrangement on this account is not rare; nor was
Davy wise enough to discern that Faraday had passed
the mere assistant stage, and become a discoverer. It
is now hard to avoid magnifying this error. But had
Faraday died or ceased to work at this time, or had his
subsequent life been devoted to money-getting, instead
of to research, would anybody now dream of ascribing
jealousy to Davy? Assuredly not. Why should he be
jealous? His reputation at this time was almost without a parallel: his glory was without a cloud. He had
added to his other discoveries that of Faraday, and
after having been his teacher for seven years, his language to him was this: ' It gives me great pleasure to
hear that you are comfortable at the Royal Institution,
and I trust that you will not only do something good
and honourable for yourself, but also for science.' This
is not the language of jealousy, potential or actual.
But the chlorine business introduced irritation and
anger, to which, and not to any ignobler motive, Davy's

opposition to the election of Faraday to the Royal Society is, I am persuaded, to be ascribed.

These matters are touched upon with perfect candour, and becoming consideration, in the volumes of Dr. Bence Jones; but in 'society' they are not always so handled. Here a name of noble intellectual associations is surrounded by injurious rumours which I would willingly scatter for ever. The pupil's magnitude, and the splendour of his position, are too great and absolute to need as a foil the humiliation of his master. Brothers in intellect, Davy and Faraday, however, could never have become brothers in feeling; their characters were too unlike. Davy loved the pomp and circumstance of fame; Faraday the inner consciousness that he had fairly won renown. They were both proud men. But with Davy pride projected itself into the outer world; while with Faraday it became a steadying and dignifying inward force. In one great particular they agreed. Each of them could have turned his science to immense commercial profit, but neither of them did so. The noble excitement of research, and the delight of discovery, constituted their reward. I commend them to the reverence which great gifts greatly exercised ought to inspire. They were both ours; and through the coming centuries England will be able to point with just pride to the possession of such men.

The first volume of the 'Life and Letters' reveals to us the youth who was to be father to the man. Skilful, aspiring, resolute, he grew steadily in knowledge and in power. Consciously or unconsciously, the relation of Action to Reaction was ever present to

Faraday's mind.   It had been fostered by his discovery
of Magnetic Rotations, and it planted in him more
daring ideas of a similar kind.   Magnetism he knew
could be evoked by electricity, and he thought that
electricity, in its turn, ought to be capable of evolution
by magnetism.   On August 29, 1831, his experiments
on this subject began.   He had been fortified by
previous trials, which, though failures, had begotten
instincts directing him towards the truth.   He, like
every strong worker, might at times miss the outward
object, but he always gained the inner light. education,
and expansion.   Of this Faraday's life was a constant
illustration.   By November he had discovered and col-
ligated a multitude of the most wonderful and unexpected phenomena.   He had generated currents by
currents; currents by magnets, permanent and transitory; and he afterwards generated currents by the
earth itself.   Arago's ' Magnetism of Rotation,' which
had for years offered itself as a challenge to the best
scientific intellects of Europe, now fell into his hands.
It proved to be a beautiful, but still special, illustration
of the great principle of Magneto-electric Induction.
Nothing equal to this latter, in the way of pure experimental enquiry, had previously been achieved.

Electricities from various sources were next examined, and their differences and resemblances revealed.
He thus assured himself of their substantial identity.
He then took up Conduction, and gave many striking
illustrations of the influence of Fusion on Conducting
Power.   Renouncing professional work, from which at
this time he might have derived an income of many
thousands a year, he poured his whole momentum into
his researches.   He was long entangled in Electrochemistry.   The light of law was for a time obscured
by the thick umbrage of novel facts; but he finally

emerged from his researches with the great principle of Definite Electro-chemical Decomposition in his hands. If his discovery of Magneto-electricity may be ranked with that of the pile by Volta, this new discovery may almost stand beside that of Definite Combining Proportions in Chemistry. He passed on to Static Electricity—its Conduction, Induction, and Mode of Propagation. He discovered and illustrated the principle of Inductive Capacity; and, turning to theory, he asked himself how electrical attractions and repulsions are transmitted. Are they, like gravity, actions at a distance, or do they require a medium? If the former, then, like gravity, they will act in straight lines; if the latter, then, like sound or light, they may turn a corner. Faraday held—and his views are gaining ground—that his experiments proved the fact of curvilinear propagation, and hence the operation of a medium. Others denied this; but none can deny the profound and philosophic character of his leading thought.[1] The first volume of the Researches contains all the papers here referred to.

Faraday had heard it stated that henceforth physical discoveries would be made solely by the aid of mathematics; that we had our data, and needed only to work deductively. Statements of a similar character crop out from time to time in our day. They arise from an imperfect acquaintance with the nature, present condition, and prospective vastness of the field ot physical enquiry. The tendency of natural science doubtless is to bring all physical phenomena under the dominion of mechanical laws; to give them, in other words, mathematical expression. But our approach to

---

[1] In a very remarkable paper published in Poggendorff's 'Annalen' for 1857, Werner Siemens accepts and develops Faraday's theory of Molecular Induction.

this result is asymptotic; and for ages to come—possibly for all the ages of the human race—Nature will find room for both the philosophical experimenter and the mathematician. Faraday entered his protest against the foregoing statement by labelling his investigations 'Experimental Researches in Electricity.' They were completed in 1854, and three volumes of them have been published. For the sake of reference, he numbered every paragraph, the last number being 3362. In 1859 he collected and published a fourth volume of papers, under the title, 'Experimental Researches in Chemistry and Physics.' Thus did this apostle of experiment illustrate its power, and magnify his office.

The second volume of the Researches embraces memoirs on the Electricity of the Gymnotus; on the Source of Power in the Voltaic Pile; on the Electricity evolved by the Friction of Water and Steam, in which the phenomena and principles of Sir William Armstrong's Hydro-electric machine are described and developed; a paper on Magnetic Rotations, and Faraday's letters in relation to the controversy it aroused. The contribution of most permanent value here, is that on the Source of Power in the Voltaic Pile. By it the Contact Theory, pure and simple, was totally overthrown, and the necessity of chemical action to the maintenance of the current demonstrated.

The third volume of the Researches opens with a memoir entitled 'The Magnetisation of Light,' and the 'Illumination of Magnetic Lines of Force.' It is difficult even now to affix a definite meaning to this title; but the discovery of the rotation of the plane of polarisation, which it announced, seems pregnant with great results. The writings of William Thomson on the theoretic aspects of the discovery; the excellent electro-dynamic measurements of Wilhelm Weber, which are models of experimental completeness and skill; Weber's labours in conjunction with his lamented friend Kohlrausch—above all, the researches of Clerk Maxwell on the Electro-magnetic Theory of Light—point to that wonderful and mysterious medium, which is the vehicle of light and radiant heat, as the probable basis also of magnetic and electric phenomena. The hope of such a connection was first raised by the discovery here referred to.[1] Faraday himself seemed to cling with particular affection to this discovery. He felt that there was more in it than he was able to unfold. He predicted that it would grow in meaning with the growth of science. This it has done; this it is doing now. Its right interpretation will probably mark an epoch in scientific history.

Rapidly following it is the discovery of Diamagnetism, or the repulsion of matter by a magnet. Brugmans had shown that bismuth repelled a magnetic needle. Here he stopped. Le Bailliff proved that antimony did the same. Here he stopped. Seebeck, Becquerel, and others, also touched the discovery. These fragmentary gleams excited a momentary curiosity and were almost forgotten, when Faraday independently alighted on the same facts; and, instead of stopping, made them the inlets to a new and vast region of

---

[1] A letter addressed to me by Professor Weber on March 18 last contains the following reference to the connection here mentioned: 'Die Hoffnung einer solchen Combination ist durch Faraday's Entdeckung der Drehung der Polarisationsebene durch magnetische Directionskraft zuerst, und sodann durch die Uebereinstimmung derjenigen Geschwindigkeit, welche das Verhältniss der electro-dynamischen Einheit zur electro-statischen ausdrückt, mit der Geschwindigkeit des Lichts angeregt worden; und mir scheint von allen Versuchen, welche zur Verwirklichung dieser Hoffnung gemacht worden sind, das von Herrn Maxwell gemachte am orfolgreichsten.'

research. The value of a discovery is to be measured by the intellectual action it calls forth; and it was Faraday's good fortune to strike such lodes of scientific truth as give occupation to some of the best intellects of our age.

The salient quality of Faraday's scientific character reveals itself from beginning to end of these volumes; a union of ardour and patience—the one prompting the attack, the other holding him on to it, till defeat was final or victory assured. Certainty in one sense or the other was necessary to his peace of mind. The right method of investigation is perhaps incommunicable; it depends on the individual rather than on the system, and the mark is missed when Faraday's researches are pointed to as merely illustrative of the power of the inductive philosophy. The brain may be filled with that philosophy; but without the energy and insight which this man possessed, and which with him were personal and distinctive, we should never rise to the level of his achievements. His power is that of individual genius, rather than of philosophic method; the energy of a strong soul expressing itself after its own fashion, and acknowledging no mediator between it and Nature.

The second volume of the 'Life and Letters,' like the first, is a historic treasury as regards Faraday's work and character, and his scientific and social relations. It contains letters from Humboldt, Herschel, Hachette, De la Rive, Dumas, Liebig, Melloni, Becquerel, Oersted, Plücker, Du Bois Reymond, Lord Melbourne, Prince Louis Napoleon, and many other distinguished men. I notice with particular pleasure a letter from Sir John Herschel, in reply to a sealed packet addressed to him by Faraday, but which he had permission to open if he pleased. The packet referred to one of the many unfulfilled hopes which spring up in the minds of fertile investigators :—

'Go on and prosper, "from strength to strength," like a victor marching with assured step to further conquests; and be certain that no voice will join more heartily in the peans that already begin to rise, and will speedily swell into a shout of triumph, astounding even to yourself, than that of J. F. W. Herschel.'

Faraday's behaviour to Melloni in 1835 merits a word of notice. The young man was a political exile in Paris. He had newly fashioned and applied the thermo-electric pile, and had obtained with it results of the greatest importance. But they were not appreciated. With the sickness of disappointed hope Melloni waited for the report of the Commissioners, appointed by the Academy of Sciences to examine the Primier. At length he published his researches in the 'Annales de Chimie.' They thus fell into the hands of Faraday, who, discerning at once their extraordinary merit, obtained for their author the Rumford Medal of the Royal Society. A sum of money always accompanies this medal; and the pecuniary help was, at this time, even more essential than the mark of honour to the young refugee. Melloni's gratitude was boundless :—

'Et vous, monsieur,' he writes to Faraday, 'qui appartenez à une société à laquelle je n'avais rien offert, vous qui me connaissiez à peine de nom ; vous n'avez pas demandé si j'avais des ennemis faibles ou puissants, ni calculé quel en était le nombre ; mais vous avez parlé pour l'opprimé étranger, pour celui qui n'avait pas le moindre droit à tant de bienveillance, et vos paroles ont été accueillies favorablement par des collègues consciencieux ! Je reconnais bien là des hommes dignes de leur noble mission, les véritable représentants de la science d'un pays libre et généreux.'

Within the prescribed limits of this article it would
be impossible to give even the slenderest summary of
Faraday's correspondence, or to carve from it more than
the merest fragments of his character. His letters,
written to Lord Melbourne and others in 1836, regarding his pension, illustrate his uncompromising independ-
ence. The Prime Minister had offended him, but
assuredly the apology demanded and given was complete. I think it certain that, notwithstanding the
very full account of this transaction given by Dr. Bence
Jones, motives and influences were at work which even
now are not entirely revealed. The minister was bitterly attacked, but he bore the censure of the press with
great dignity. Faraday, while he disavowed having
either directly or indirectly furnished the matter of
those attacks, did not publicly exonerate the Premier.
The Hon. Caroline Fox had proved herself Faraday's
ardent friend, and it was she who had healed the breach
between the philosopher and the minister. She mani
festly thought that Faraday ought to have come forward in Lord Melbourne's defence, and there is a flavour
of resentment in one of her letters to him on the subject. No doubt Faraday had good grounds for his
reticence, but they are to me unknown.

In 1841 his health broke down utterly, and he went
to Switzerland with his wife and brother-in-law. His
bodily vigour soon revived, and he accomplished feats
of walking respectable even for a trained mountaineer.
The published extracts from his Swiss journal contain
many beautiful and touching allusions. Amid references
to the tints of the Jungfrau, the blue rifts of the glaciers,
and the noble Niesen towering over the Lake of Thun,
we come upon the charming little scrap which I have
elsewhere quoted : 'Clout-nail making goes on here
rather considerably, and is a very neat and pretty

operation to observe. I love a smith's shop and anything relating to smithery. My father was a smith.'
This is from his journal; but he is unconsciously speak·
ing to somebody—perhaps to the world.

His description of the Staubbach, Giessbach, and
of the scenic effects of sky. and mountain, are all fine
and sympathetic. But amid it all, and in reference to
it all, he tells his sister that 'true enjoyment is from
within, not from without.' In those days Agassiz was
living under a slab of gneiss on the glacier of the Aar.
Faraday met Forbes at the Grimsel, and arranged
with him an excursion to the 'Hôtel des Neuchâtelois';
but indisposition put the project out.

From the Fort of Ham, in 1843, Faraday received
a letter addressed to him by Prince Louis Napoleon
Bonaparte. He read this letter to me many years ago,
and the desire, shown in various ways by the French
Emperor, to turn modern science to account, has often
reminded me of it since. At the age of thirty-five the
prisoner of Ham speaks of 'rendering his captivity
less sad by studying the great discoveries' which
science owes to Faraday; and he asks a question which
reveals his cast of thought at the time : ' What is the
most simple combination to give to a voltaic battery,
in order to produce a spark capable of setting fire to
powder under water or under ground?' Should the
necessity arise, the French Emperor will not lack at
the outset the best appliances of modern science; while
we, I fear, shall have to learn the magnitude of the
resources we are now neglecting amid the pangs of
actual war.[1]

[1] The 'science' has since been applied, with astonishing effect,
by those who had studied it far more thoroughly than the Emperor
of the French. We also, I am happy to think, have improved the
time since the above words were written [1873].

One turns with renewed pleasure to Faraday's letters to his wife, published in the second volume. Here surely the loving essence of the man appears more distinctly than anywhere else. From the house of Dr. Percy, in Birmingham, he writes thus :—

'Here—even here—the moment I leave the table, I wish I were with you IN QUIET. Oh, what happiness is ours ! My runs into the world in this way only serve to make me esteem that happiness the more.'

And again :

'We have been to a grand conversazione in the town-hall, and I have now returned to my room to talk with you, as the pleasantest and happiest thing that I can do. Nothing rests me so much as communion with you. I feel it even now as I write, and catch myself saying the words aloud as I write them.'

Take this, moreover, as indicative of his love for Nature :

'After writing, I walk out in the evening hand in hand with my dear wife to enjoy the sunset ; for to me who love scenery, of all that I have seen or can see, there is none surpasses that of heaven. A glorious sunset brings with it a thousand thoughts that delight me.'

Of the numberless lights thrown upon him by the 'Life and Letters,' some fall upon his religion. In a letter to Lady Lovelace, he describes himself as belonging to 'a very small and despised sect of Christians, known, if known at all, as *Sandemanians*, and our hope is founded on the faith that is in Christ.' He adds : ' I do not think it at all necessary to tie the study of the natural sciences and religion together, and in my intercourse with my fellow-creatures, that which is religious, and that which is philosophical, have ever been two distinct things.' He saw clearly the danger of quitting his moorings, and his science acted indirectly as the safeguard of his faith. For his investigations so filled his mind as to leave no room for sceptical questionings, thus shielding from the assaults of philosophy the creed of his youth. His religion was constitutional and hereditary. It was implied in the eddies of his blood and in the tremors of his brain; and, however its outward and visible form might have changed, Faraday would still have possessed its elemental constituents—awe, reverence, truth, and love.

It is worth enquiring how so profoundly religious a mind, and so great a teacher, would be likely to regard our present discussions on the subject of education. Faraday would be a 'secularist' were he now alive. He had no sympathy with those who contemn knowledge unless it be accompanied by dogma. A lecture delivered before the City Philosophical Society in 1818, when he was twenty-six years of age, expresses the views regarding education which he entertained to the end of his life. 'First, then,' he says, 'all theological considerations are banished from the society, and of course from my remarks; and whatever I may say has no reference to a future state, or to the means which are to be adopted in this world in anticipation of it. Next, I have no intention of substituting anything for religion, but I wish to take that part of human nature which is independent of it. Morality, philosophy, commerce, the various institutions and habits of society, are independent of religion, and may exist either with or without it. They are always the same, and can dwell alike in the breasts of those who, from opinion, are entirely opposed in the set of principles they include in the term religion, or in those who have none.

'To discriminate more closely, if possible, I will observe that we have *no* right to judge religious opinions; but the human nature of this evening is that part of man which we *have* a right to judge. And I think it will be found on examination, that this humanity—as it may perhaps be called—will accord with what I have before described as being in our own hands so improvable and perfectible.'

In an old journal I find the following remarks on one of my earliest dinners with Faraday: 'At two o'clock he came down for me. He, his niece, and myself, formed the party, " I never give dinners," he said. " I don't know how to give dinners, and I never dine out. But I should not like my friends to attribute this to a wrong cause. I act thus for the sake of securing time for work, and not through religious motives, as some imagine." He said grace. I am almost ashamed to call his prayer a " saying of grace." In the language of Scripture, it might be described as the petition of a son, into whose heart God had sent the Spirit of His Son, and who with absolute trust asked a blessing from his father. We dined on roast beef, Yorkshire pudding, and potatoes; drank sherry, talked of research and its requirements, and of his habit of keeping himself free from the distractions of society. He was bright and joyful—boylike, in fact, though he is now sixty-two. His work excites admiration, but contact with him warms and elevates the heart. Here, surely, is a strong man. I love strength; but let me not forget the example of its union with modesty, tenderness, and sweetness, in the character of Faraday.'

Faraday's progress in discovery, and the salient points of his character, are well brought out by the wise choice of letters and extracts published in the volumes before us. I will not call the labours of the biographer final. So great a character will challenge reconstruction. In the coming time some sympathetic spirit, with the requisite strength, knowledge, and solvent power, will, I doubt not, render these materials plastic, give them more perfect organic form, and send through them, with less of interruption, the currents of Faraday's life. 'He was too good a man,' writes his present biographer, 'for me to estimate rightly, and too great a philosopher for me to understand thoroughly.' That may be: but the reverent affection to which we owe the discovery, selection, and arrangement of the materials here placed before us, is probably a surer guide than mere literary skill. The task of the artist who may wish in future times to reproduce the real though unobtrusive grandeur, the purity, beauty, and childlike simplicity of him whom we have lost, will find his chief treasury already provided for him by Dr. Bence Jones's labour of love.

XIX.

THE COPLEY MEDALIST OF 1870.

THIRTY years ago Electro-magnetism was looked to
as a motive power, which might possibly compete with steam.   In centres of industry, such as
Manchester, attempts to investigate and apply this
power were numerous.   This is shown by the scientific
literature of the time.   Among others Mr. James
Prescot Joule, a resident of Manchester, took up the
subject, and, in a series of papers published in Stur-
geon's 'Annals of Electricity' between 1839 and 1841,
described various attempts at the construction and perfection of electro-magnetic engines.   The spirit in
which Mr. Joule pursued these enquiries is revealed in
the following extract : 'I am particularly anxious,' he
says, 'to communicate any new arrangement in order,
if possible, to forestall the monopolising designs of those
who seem to regard this most interesting subject merely
in the light of pecuniary speculation.'   He was naturally led to investigate the laws of electro-magnetic
attractions, and in 1840 he announced the important
principle that the attractive force exerted by two electro-
magnets, or by an electro-magnet and a mass of annealed iron, is directly proportional to the square
of the strength of the magnetising current; while
the attraction exerted between an electro-magnet
and the pole of a permanent steel magnet, varies

simply as the strength of the current. These investigations were conducted independently of, though a little subsequently to, the celebrated enquiries of Henry, Jacobi, and Lenz and Jacobi, on the same subject.

On December 17, 1840, Mr. Joule communicated to the Royal Society a paper on the production of heat by Voltaic electricity. In it he announced the law that the calorific effects of equal quantities of transmitted electricity are proportional to the resistance overcome by the current, whatever may be the length, thickness, shape, or character of the metal which closes the circuit ; and also proportional to the square of the quantity of transmitted electricity. This is a law of primary importance. In another paper, presented to, but declined by, the Royal Society, he confirmed this law by new experiments, and materially extended it. He also executed experiments on the heat consequent on the passage of Voltaic electricity through electrolytes, and found, in all cases, that the heat evolved by the proper action of any Voltaic current is proportional to the square of the intensity of that current, multiplied by the resistance to conduction which it experiences. From this law he deduced a number of conclusions of the highest importance to electrochemistry.

It was during these enquiries, which are marked throughout by rare sagacity and originality, that the great idea of establishing quantitative relations between Mechanical Energy and Heat arose and assumed definite form in his mind. In 1843 Mr. Joule read before the meeting of the British Association at Cork a paper 'On the Calorific Effects of Magneto-Electricity, and on the Mechanical Value of Heat.' Even at the present day this memoir is tough reading, and at the time it was

written it must have appeared hopelessly entangled.
This, I should think, was the reason why Faraday
advised Mr. Joule not to submit the paper to the
Royal Society. But its drift and results are summed
up in these memorable words by its author, written
some time subsequently: 'In that paper it was demonstrated experimentally, that the mechanical power
exerted in turning a magneto-electric machine is converted into the heat evolved by the passage of the
currents of induction through its coils; and, on the
other hand, that the motive power of the electro-
magnetic engine is obtained at the expense of the heat
due to the chemical reaction of the battery by which it
is worked.'[1] It is needless to dwell upon the weight
and importance of this statement.

Considering the imperfections incidental to a first
determination, it is not surprising that the 'mechanical values of heat,' deduced from the different series
of experiments published in 1843, varied widely
from each other. The lowest limit was 587, and
the highest 1,026 foot-pounds, for 1° Fahr. of temperature.

One noteworthy result of his enquiries, which was
pointed out at the time by Mr. Joule, had reference to
the exceedingly small fraction of the heat actually
converted into useful effect in the steam-engine. The
thoughts of the celebrated Julius Robert Mayer, who
was then engaged in Germany upon the same question,
had moved independently in the same groove; but to
his labours due reference will be made on a future
occasion.[2] In the memoir now referred to, Mr. Joule
also announced that he had proved heat to be evolved
during the passage of water through narrow tubes; and

[1] Phil. Mag., May 1845.   [2] See the next Fragment.

he deduced from these experiments an equivalent of
770 foot-pounds, a figure remarkably near the one now
accepted. A detached statement regarding the origin
and convertibility of animal heat strikingly illustrates
the penetration of Mr. Joule, and his mastery of principles, at the period now referred to. A friend had
mentioned to him Haller's hypothesis, that animal heat
might arise from the friction of the blood in the veins
and arteries. 'It is unquestionable,' writes Mr. Joule,
'that heat is produced by such friction ; but it must be
understood that the mechanical force expended in the
friction is a part of the force of affinity which causes
the venous blood to unite with oxygen, so that the
whole heat of the system must still be referred to the
chemical changes. But if the animal were engaged in
turning a piece of machinery, or in ascending a mountain, I apprehend that in proportion to the muscular
effort put forth for the purpose, a *diminution* of the
heat evolved in the system by a given chemical action
would be experienced.' The italics in this memorable
passage, written, it is to be remembered, in 1843, are
Mr. Joule's own.

The concluding paragraph of this British Association
paper equally illustrates his insight and precision,
regarding the nature of chemical and latent heat. 'I
had,' he writes, 'endeavoured to prove that when two
atoms combine together, the heat evolved is exactly
that which would have been evolved by the electrical
current due to the chemical action taking place, and is
therefore proportional to the intensity of the chemical
force causing the atoms to combine. I now venture to
state more explicitly, that it is not precisely the attraction of affinity, but rather the mechanical force expended by the atoms in falling towards one another,

which determines the intensity of the current, and, consequently, the quantity of heat evolved; so that we have a simple hypothesis by which we may explain why heat is evolved so freely in the combination of gases, and by which indeed we may account " latent heat " as a mechanical power, prepared for action, as a watch-spring is when wound up. Suppose, for the sake of illustration, that 8 lbs. of oxygen and 1 lb. of hydrogen were presented to one another in the gaseous state, and then exploded; the heat evolved would be about 1° Fahr. in 60,000 lbs. of water, indicating a mechanical force, expended in the combination, equal to a weight of about 50,000,000 lbs. raised to the height of one foot. Now if the oxygen and hydrogen could be presented to each other in a liquid state, the heat of combination would be less than before, because the atoms in combining would fall through less space.' No words of mine are needed to point out the commanding grasp of molecular physics, in their relation to the mechanical theory of heat, implied by this statement.

Perfectly assured of the importance of the principle which his experiments aimed at establishing, Mr. Joule did not rest content with results presenting such discrepancies as those above referred to. He resorted in 1844 to entirely new methods, and made elaborate experiments on the thermal changes produced in air during its expansion : firstly, against a pressure, and therefore performing work ; secondly, against no pressure, and therefore performing no work. He thus established anew the relation between the heat consumed and the work done. From five different series of experiments he deduced five different mechanical equivalents ; the agreement between them being far greater than that

attained in his first experiments. The mean of them
was 802 foot-pounds. From experiments with water
agitated by a paddle-wheel, he deduced, in 1845, an
equivalent of 890 foot-pounds. In 1847 he again
operated upon water and sperm-oil, agitated them by a
paddle-wheel, determined their elevation of temperature,
and the mechanical power which produced it. From
the one he derived an equivalent of 781·5 foot-pounds;
from the other an equivalent of 782·1 foot-pounds.
The mean of these two very close determinations is
781·8 foot-pounds.

By this time the labours of the previous ten years
had made Mr. Joule completely master of the conditions
essential to accuracy and success. Bringing his ripened
experience to bear upon the subject, he executed in
1849 a series of 40 experiments on the friction of water,
50 experiments on the friction of mercury, and 20
experiments on the friction of plates of cast-iron. He
deduced from these experiments our present mechanical
equivalent of heat, justly recognised all over the world
as ' Joule's equivalent.'

There are labours so great and so pregnant in consequences, that they are most highly praised when they
are most simply stated. Such are the labours of Mr.
Joule. They constitute the experimental foundation of
a principle of incalculable moment, not only to the
practice, but still more to the philosophy of Science.
Since the days of Newton, nothing more important than
the theory, of which Mr. Joule is the experimental
demonstrator, has been enunciated.

1 have omitted all reference to the numerous minor
papers with which Mr. Joule has enriched scientific
literature. Nor have I alluded to the important investigations which he has conducted jointly with Sir

William Thomson.   But sufficient, I think, has been
here said to show that, in conferring upon Mr. Joule
the highest honour of the Royal Society, the Council
paid to genius not only a well-won tribute, but one
which had been fairly earned twenty years previously.[1]

[1] Lord Beaconsfield has recently honoured himself and England
by bestowing an annual pension of 200*l.* on Dr. Joule.

XX

THE COPLEY MEDALIST OF 1871.

DR. JULIUS ROBERT MAYER was educated for the medical profession. In the summer of 1840, as he himself informs us, he was at Java, and there observed that the venous blood of some of his patients had a singularly bright red colour. The observation riveted his attention; he reasoned upon it, and came to the conclusion that the brightness of the colour was due to the fact that a less amount of oxidation sufficed to keep up the temperature of the body in a hot climate than in a cold one. The darkness of the venous blood he regarded as the visible sign of the energy of the oxidation.

It would be trivial to remark that accidents such as this, appealing to minds prepared for them, have often led to great discoveries. Mayer's attention was thereby drawn to the whole question of animal heat. Lavoisier had ascribed this heat to the oxidation of the food. 'One great principle,' says Mayer, 'of the physiological theory of combustion, is that under all circumstances the same amount of fuel yields, by its perfect combustion, the same amount of heat; that this law holds good even for vital processes; and that hence the living body, notwithstanding all its enigmas and wonders, is incompetent to generate heat out of nothing.'

But beyond the power of generating internal heat,

the animal organism can also generate heat outside of itself. A blacksmith, for example, by hammering can heat a nail, and a savage by friction can warm wood to its point of ignition. Now, unless we give up the physiological axiom that the living body cannot create heat out of nothing, 'we are driven,' says Mayer, 'to the conclusion that it is the *total* heat generated within and *without* that is to be regarded as the true calorific effect of the matter oxidised in the body.'

From this, again, he inferred that the heat generated externally must stand in a fixed relation to the work expended in its production. For, supposing the organic processes to remain the same ; if it were possible, by the mere alteration of the apparatus, to generate different amounts of heat by the same amount of work, it would follow that the oxidation of the same amount of material would sometimes yield a less, sometimes a greater, quantity of heat. 'Hence,' says Mayer, 'that a fixed relation subsists between heat and work, is a postulate of the physiological theory of combustion.'

This is the simple and natural account, given subsequently by Mayer himself, of the course of thought started by his observation in Java. But the conviction once formed, that an unalterable relation subsists between work and heat, it was inevitable that Mayer should seek to express it numerically. It was also inevitable that a mind like his, having raised itself to clearness on this important point, should push forward to consider the relationship of natural forces generally. At the beginning of 1842 his work had made considerable progress ; but he had become physician to the town of Heilbronn, and the duties of his profession limited the time which he could devote to purely scientific enquiry. He thought it wise, therefore, to secure himself against accident, and in the spring of 1842 wrote to Liebig, asking him to publish in his 'Annalen' a brief preliminary notice of the work then accomplished. Liebig did so, and Dr. Mayer's first paper is contained in the May number of the 'Annalen' for 1842.

Mayer had reached his conclusions by reflecting on the complex processes of the living body; but his first step in public was to state definitely the physical principles on which his physiological deductions were to rest. He begins, therefore, with the forces of inorganic nature. He finds in the universe two systems of causes which are not mutually convertible;—the different kinds of matter and the different forms of force. The first quality of both he affirms to be *indestructibility*. A force cannot become nothing, nor can it arise from nothing. Forces are convertible but not destructible. In the terminology of his time, he then gives clear expression to the ideas of potential and dynamic energy, illustrating his point by a weight resting upon the earth, suspended at a height above the earth, and actually falling to the earth. He next fixes his attention on cases where motion is apparently destroyed, without producing other motion; on the shock of inelastic bodies, for example. Under what form does the vanished motion maintain itself? Experiment alone, says Mayer, can help us here. He warms water by stirring it; he refers to the force expended in overcoming friction. Motion in both cases disappears; but heat is generated, and the quantity generated is the equivalent of the motion destroyed. 'Our locomotives,' he observes with extraordinary sagacity, 'may be compared to distilling apparatus: the heat beneath the boiler passes into the motion of the train, and is again deposited as heat in the axles and wheels.'

A numerical solution of the relation between heat and work was what Mayer aimed at, and towards the end of his first paper he makes the attempt. It was known that a definite amount of air, in rising one degree in temperature, can take up two different amounts of heat. If its volume be kept constant, it takes up one amount: if its pressure be kept constant it takes up a different amount. These two amounts are called the specific heat under constant volume and under constant pressure. The ratio of the first to the second is as $1 : 1\cdot421$. No man, to my knowledge, prior to Dr. Mayer, penetrated the significance of these two numbers. He first saw that the excess $0\cdot421$ was not, as then universally supposed, heat actually lodged in the gas, but heat which had been actually consumed by the gas in expanding against pressure. The amount of work here performed was accurately known, the amount of heat consumed was also accurately known, and from these data Mayer determined the mechanical equivalent of heat. Even in this first paper he is able to direct attention to the enormous discrepancy between the theoretic power of the fuel consumed in steam-engines, and their useful effect.

Though this paper contains but the germ of his further labours, I think it may be safely assumed that, as regards the mechanical theory of heat, this obscure Heilbronn physician, in the year 1842, was in advance of all the scientific men of the time.

Having, by the publication of this paper, secured himself against what he calls 'Eventualitäten,' he devoted every hour of his spare time to his studies, and in 1845 published a memoir which far transcends his first one in weight and fulness, and, indeed, marks an epoch in the history of science. The title of Mayer's first paper was, ' Remarks on the Forces of Inorganic Nature.' The title of his second great essay was, 'Organic Motion in its Connection with Nutrition.' In it he expands and illustrates the physical principles laid down in his first

brief paper. He goes fully through the calculation of
the mechanical equivalent of heat. He calculates the
performances of steam-engines, and finds that 100 lbs.
of coal, in a good working engine, produce only the
same amount of heat as 95 lbs. in an unworking one;
the 5 missing lbs. having been converted into work.
He determines the useful effect of gunpowder, and finds
nine per cent. of the force of the consumed charcoal invested on the moving ball. He records observations on
the heat generated in water agitated by the pulping-
engine of a paper manufactory, and calculates the equivalent of that heat in horse-power. He compares
chemical combination with mechanical combination—
the union of atoms with the union of falling bodies with
the earth. He calculates the velocity with which a
body starting at an infinite distance would strike the
earth's surface, and finds that the heat generated by its
collision would raise an equal weight of water 17,356°
C. in temperature. He then determines the thermal
effect which would be produced by the earth itself falling
into the sun. So that here, in 1845, we have the germ
of that meteoric theory of the sun's heat which Mayer
developed with such extraordinary ability three years
afterwards. He also points to the almost exclusive effi-
cacy of the sun's heat in producing mechanical motions
upon the earth, winding up with the profound remark,
that the heat developed by friction in the wheels of our
wind and water mills comes from the sun in the form
of vibratory motion; while the heat produced by mills
driven by tidal action is generated at the expense of the
earth's axial rotation.

Having thus, with firm step, passed through the
powers of inorganic nature, his next object is to bring
his principles to bear upon the phenomena of vegetable
and animal life. Wood and coal can burn; whence

come their heat, and the work producible by that heat? From the immeasurable reservoir of the sun. Nature has proposed to herself the task of storing up the light which streams earthward from the sun, and of casting into a permanent form the most fugitive of all powers. To this end she has overspread the earth with organisms which, while living, take in the solar light, and by its consumption generate forces of another kind. These organisms are plants. The vegetable world, indeed, constitutes the instrument whereby the wave-motion of the sun is changed into the rigid form of chemical tension, and thus prepared for future use. With this prevision, as shall subsequently be shown, the existence of the human race itself is inseparably connected. It is to be observed that Mayer's utterances are far from being anticipated by vague statements regarding the 'stimulus' of light, or regarding coal as 'bottled sunlight.' He first saw the full meaning of De Saussure's observation as to the reducing power of the solar rays, and gave that observation its proper place in the doctrine of conservation. In the leaves of a tree, the carbon and oxygen of carbonic acid, and the hydrogen and oxygen of water, are forced asunder at the expense of the sun, and the amount of power thus sacrificed is accurately restored by the combustion of the tree. The heat and work potential in our coal strata are so much strength withdrawn from the sun of former ages. Mayer lays the axe to the root of the notions regarding 'vital force' which were prevalent when he wrote. With the plain fact before us that in the absence of the solar rays plants cannot perform the work of reduction, or generate chemical tensions, it is, he contends, incredible that these tensions should be caused by the mystic play of the vital force. Such an hypothesis would cut off all investigation; it would land us in a chaos of unbridled phantasy.

' I count,' he says, ' therefore, upon your agreement with
me when I state, as an axiomatic truth, that during
vital processes the *conversion* only, and never the
creation of matter or force occurs.'

Having cleared his way through the vegetable world,
as he had previously done through inorganic nature,
Mayer passes on to the other organic kingdom.   The
physical forces collected by plants become the property
of animals.   Animals consume vegetables, and cause
them to reunite with the atmospheric oxygen.   Animal
heat is thus produced ; and not only animal heat, but
animal motion.   There is no indistinctness about Mayer
here; he grasps his subject in all its details, and reduces
to figures the concomitants of muscular action.   A
bowler who imparts to an 8-lb. ball a velocity of 30 feet,
consumes in the act $\frac{1}{10}$ of a grain of carbon.   A man
weighing 150 lbs., who lifts his own body to a height
of 8 feet, consumes in the act 1 grain of carbon.   In
climbing a mountain 10,000 feet high, the consumption
of the same man would be 2 oz. 4 drs. 50 grs. of carbon.
Boussingault had determined experimentally the addition to be made to the food of horses when actively
working, and Liebig had determined the addition to be
made to the food of men.   Employing the mechanical
equivalent of heat, which he had previously calculated,
Mayer proves the additional food to be amply sufficient
to cover the increased oxidation.

But he does not content himself with showing, in a
general way, that the human body burns according to
definite laws, when it performs mechanical work.   He
seeks to determine the particular portion of the body
consumed, and in doing so executes some noteworthy
calculations.   The muscles of a labourer 150 lbs. in
weight weigh 64 lbs.; but when perfectly desiccated they
fall to 15 lbs.   Were the oxidation corresponding to

that labourer's work exerted on the muscles alone, they would be utterly consumed in 80 days. The heart furnishes a still more striking example. Were the oxidation necessary to sustain the heart's action exerted upon its own tissue, it would be utterly consumed in 8 days. And if we confine our attention to the two ventricles, their action would be sufficient to consume the associated muscular tissue in $3\frac{1}{2}$ days. Here, in his own words, emphasised in his own way, is Mayer's pregnant conclusion from these calculations: 'The muscle is only the apparatus by means of which the conversion of the force is effected; *but it is not the substance consumed in the production of the mechanical effect.*' He calls the blood ' the oil of the lamp of life;' it is the slow-burning fluid whose chemical force, in the furnace of the capillaries, is sacrificed to produce animal motion. This was Mayer's conclusion twenty-six years ago. It was in complete opposition to the scientific conclusions of his time; but eminent investigators have since amply verified it.

Thus, in baldest outline, I have sought to give some notion of the first half of this marvellous essay. The second half is so exclusively physiological that I do not wish to meddle with it. I will only add the illustration employed by Mayer to explain the action of the nerves upon the muscles. As an engineer, by the motion of his finger in opening a valve or loosing a detent, can liberate an amount of mechanical motion almost infinite compared with its exciting cause, so the nerves, acting upon the muscles, can unlock an amount of activity, wholly out of proportion to the work done by the nerves themselves.

As regards these questions of weightiest import to the science of physiology, Dr. Mayer, in 1845, was assuredly far in advance of all living men.

Mayer grasped the mechanical theory of heat with commanding power, illustrating it and applying it in the most diverse domains. He began, as we have seen, with physical principles ; he determined the numerical relation between heat and work; he revealed the source of the energies of the vegetable world, and showed the relationship of the heat of our fires to solar heat. He followed the energies which were potential in the vegetable, up to their local exhaustion in the animal. But in 1845 a new thought was forced upon him by his calculations. He then, for the first time, drew attention to the astounding amount of heat generated by gravity where the force has sufficient distance to act through. He proved, as I have before stated, the heat of collision of a body falling from an infinite distance to the earth, to be sufficient to raise the temperature of a quantity of water, equal to the falling body in weight, 17,356° C. He also found, in 1845, that the gravitating force between the earth and sun was competent to generate an amount of heat equal to that obtainable from the combustion of 6,000 times the weight of the earth of solid coal. With the quickness of genius he saw that we had here a power sufficient to produce the enormous temperature of the sun, and also to account for the primal molten condition of our own planet. Mayer shows the utter inadequacy of chemical forces, as we know them, to produce or maintain the solar temperature. He shows that were the sun a lump of coal it would be utterly consumed in 5,000 years. He shows the difficulties attending the assumption that the sun is a cooling body; for, supposing it to possess even the high specific heat of water, its temperature would fall 15,000° in 5,000 years. He finally concludes that the light and heat of the sun are maintained by the constant impact of meteoric matter. I never ventured an opinion as to the truth of this theory; that is a question which may still have to be fought out. But I refer to it as an illustration of the force of genius with which Mayer followed the mechanical theory of heat through all its applications. Whether the meteoric theory be a matter of fact or not, with him abides the honour of proving to demonstration that the light and heat of suns and stars *may* be originated and maintained by the collisions of cold planetary matter.

It is the man who with the scantiest data could accomplish all this in six. short years, and in the hours snatched from the duties of an arduous profession, that the Royal Society, in 1871, crowned with its highest honour.

Comparing this brief history with that of the Copley Medalist of 1870, the differentiating influence of 'environment,' on two minds of similar natural cast and endowment, comes out in an instructive manner. Withdrawn from mechanical appliances, Mayer fell back upon reflection, selecting with marvellous sagacity, from existing physical data, the single result on which could be founded a calculation of the mechanical equivalent of heat. In the midst of mechanical appliances, Joule resorted to experiment, and laid the broad and firm foundation which has secured for the mechanical theory the acceptance it now enjoys. A great portion of Joule's time was occupied in actual manipulation; freed from this, Mayer had time to follow the theory into its most abstruse and impressive applications. With their places reversed, however, Joule might have become Mayer, and Mayer might have become Joule.

It does not lie within the scope of these brief articles to enter upon the developments of the Dynamical Theory accomplished since Joule and Mayer executed their memorable labours.

XXI.

DEATH BY LIGHTNING.

PEOPLE in general imagine, when they think at all
about the matter, that an impression upon the
nerves—a blow, for example, or the prick of a pin—is
felt at the moment it is inflicted. But this is not the case.
The seat of sensation being the brain, to it the intelligence of any impression made upon the nerves has to be
transmitted before this impression can become manifest
as consciousness. The transmission, moreover, requires
time, and the consequence is, that a wound inflicted on
a portion of the body distant from the brain is more
tardily appreciated than one inflicted adjacent to the
brain. By an extremely ingenious experimental arrangement, Helmholtz has determined the velocity of this
nervous transmission, and finds it to be about eighty
feet a second, or less than one-thirteenth of the velocity
of sound in air. If therefore, a whale forty feet long
were wounded in the tail, it would not be conscious of
the injury till half a second after the wound had been
inflicted.[1] But this is not the only ingredient in the
delay. There can scarcely be a doubt that to every act
of consciousness belongs a determinate molecular arrangement of the brain—that every thought or feeling has

[1] A most admirable lecture on the velocity of nervous transmission has been published by Dr. Du Bois Reymond in the 'Proceedings of the Royal Institution' for 1866, vol. iv. p. 575.

29

its physical correlative in that organ; and nothing can be more certain than that every physical change, whether molecular or mechanical, requires time for its accomplishment. So that, besides the interval of transmission, a still further time is necessary for the brain to put itself in order—for its molecules to take up the motions or positions necessary to the completion of consciousness. Helmholtz considers that one-tenth of a second is demanded for this purpose. Thus, in the case of the whale above supposed, we have first half a second consumed in the transmission of the intelligence through the sensor nerves to the head, one-tenth of a second consumed by the brain in completing the arrangements necessary to consciousness, and, if the velocity of transmission through the motor be the same as that through the sensor nerves, half a second in sending a command to the tail to defend itself. Thus one second and a tenth would elapse before an impression made upon its caudal nerves could be responded to by a whale forty feet long.

Now, it is quite conceivable that an injury might be inflicted so rapidly that within the time required by the brain to complete the arrangements necessary to consciousness, its power of arrangement might be destroyed. In such a case, though the injury might be of a nature to cause death, this would occur without pain, Death in this case would be simply the sudden negation of life, without any intervention of consciousness whatever.

The time required for a rifle-bullet to pass clean through a man's head may be roughly estimated at a thousandth of a second. Here, therefore, we should have no room for sensation, and death would be painless. But there are other actions which far transcend in rapidity that of the rifle-bullet. A flash of light-

ning cleaves a cloud, appearing and disappearing in
less than a hundred-thousandth of a second, and the
velocity of electricity is such as would carry it in a
single second over a distance almost equal to that which
separates the earth and moon.   It is well known that a
luminous impression once made upon the retina endures
for about one-sixth of a second, and that this is the
reason why we see a continuous band of light when a
glowing coal is caused to pass rapidly through the air.
A body illuminated by an instantaneous flash continues
to be seen for the sixth of a second after the flash has
become extinct ; and if the body thus illuminated be in
motion, it appears at rest at the place where the flash
falls upon it. When a colour-top with differently-coloured
sectors is caused to spin rapidly the colours blend together.
Such a top, rotating in a dark room and illuminated
by an electric spark, appears motionless, each distinct
colour being clearly seen.   Professor Dove has found
that a flash of lightning produces the same effect.
During a thunderstorm he put a colour-top in exceedingly rapid motion, and found that every flash revealed
the top as a motionless object with its colours distinct.
If illuminated solely by a flash of lightning, the motion
of all bodies on the earth's surface would, as Dove has
remarked, appear suspended.   A cannon-ball, for example, would have its flight apparently arrested, and
would seem to hang motionless in space as long as the
luminous impression which revealed the ball remained
upon the eye.

If, then, a rifle-bullet move with sufficient rapidity
to destroy life without the interposition of sensation,
much more is a flash of lightning competent to produce
this effect.   Accordingly, we have well-authenticated
cases of people being struck senseless by lightning who,

on recovery, had no memory of pain. The following
circumstantial case is described by Hemmer :—

On June 30, 1788, a soldier in the neighbourhood
of Mannheim, being overtaken by rain, placed himself
under a tree, beneath which a woman had previously
taken shelter. He looked upwards to see whether the
branches were thick enough to afford the required protection, and, in doing so, was struck by lightning, and
fell senseless to the earth. The woman at his side experienced the shock in her foot, but was not struck
down. Some hours afterwards the man revived, but
remembered nothing about what had occurred, save the
fact of his looking up at the branches. This was his
last act of consciousness, and he passed from the conscious to the unconscious condition without pain. The
visible marks of a lightning stroke are usually insignificant : the hair is sometimes burnt ; slight wounds are
observed ; while, in some instances, a red streak marks
the track of the discharge over the skin.

Under ordinary circumstances, the discharge from
a small Leyden jar is exceedingly unpleasant to me.
Some time ago I happened to stand in the presence of
a numerous audience, with a battery of fifteen large
Leyden jars charged beside me. Through some awk-
wardness on my part, I touched a wire leading from the
battery, and the discharge went through my body.
Life was absolutely blotted out for a very sensible
interval, without a trace of pain. In a second or so consciousness returned ; I vaguely discerned the audience
and apparatus, and, by the help of these external
appearances, immediately concluded that I had received
the battery discharge. The *intellectual* consciousness
of my position was restored with exceeding rapidity,
but not so the *optical* consciousness. To prevent the
audience from being alarmed, I observed that it had

often been my desire to receive accidentally such a
. shock, and that my wish had at length been fulfilled.
But, while making this remark, the appearance which
my body presented to my eyes was that of a number
of separate pieces. The arms, for example, were
detached from the trunk, and seemed suspended in the
air. In fact, memory and the power of reasoning
appeared to be complete long before the optic nerve
was restored to healthy action. But what I wish chiefly
to dwell upon here is, the absolute painlessness of the
shock; and there cannot, I think, be a doubt that, to a
person struck dead by lightning, the passage from life
to death occurs without consciousness being in the least
degree implicated. It is an abrupt stoppage of sensation, unaccompanied by a pang.

XXII.

SCIENCE AND THE 'SPIRITS.'

THEIR refusal to investigate 'spiritual phenomena' is often urged as a reproach against scientific men. I here propose to give a sketch of an attempt to apply to the 'phenomena' those methods of enquiry which are found available in dealing with natural truth.

Some years ago, when the spirits were particularly active in this country, Faraday was invited, or rather entreated, by one of his friends to meet and question them. He had, however, already made their acquaintance, and did not wish to renew it. I had not been so privileged, and he therefore kindly arranged a transfer of the invitation to me. The spirits themselves named the time of meeting, and I was conducted to the place at the day and hour appointed.

Absolute unbelief in the facts was by no means my condition of mind. On the contrary, I thought it probable that some physical principle, not evident to the spiritualists themselves, might underlie their manifestations. Extraordinary effects are produced by the accumulation of small impulses. Galileo set a heavy pendulum in motion by the well-timed puffs of his breath. Ellicot set one clock going by the ticks of another, even when the two clocks were separated by a wall. Preconceived notions can, moreover, vitiate, to an extraordinary degree, the testimony of even veracious persons. Hence my desire to witness those extraordinary phenomena, the existence of which seemed placed beyond a doubt by the known veracity of those who had witnessed and described them. The meeting took place at a private residence in the neighbourhood of London. My host, his intelligent wife, and a gentleman who may be called X., were in the house when I arrived. I was informed that the 'medium' had not yet made her appearance; that she was sensitive, and might resent suspicion. It was therefore requested that the tables and chairs should be examined before her arrival, in order to be assured that there was no trickery in the furniture. This was done; and I then first learned that my hospitable host had arranged that the *séance* should be a dinner-party. This was to me an unusual form of investigation; but I accepted it, as one of the accidents of the occasion.

The 'medium' arrived—a delicate-looking young lady, who appeared to have suffered much from ill-health. I took her to dinner and sat close beside her. Facts were absent for a considerable time, a series of very wonderful narratives supplying their place. The duty of belief on the testimony of witnesses was frequently insisted on. X. appeared to be a chosen spiritual agent, and told us many surprising things. He affirmed that, when he took a pen in his hand, an influence ran from his shoulder downwards, and impelled him to write oracular sentences. I listened for a time, offering no observation. 'And now,' continued X., 'this power has so risen as to reveal to me the thoughts of others. Only this morning I told a friend what he was thinking of, and what he intended to do during the day.' Here, I thought, is something that can be at once tested. I said immediately to X.: 'If you wish to win to your cause an apostle, who will proclaim your principles to the world from the housetop, tell me what I am now thinking of.' X. reddened, and did *not* tell me my thought.

Some time previously I had visited Baron Reichenbach, in Vienna, and I now asked the young lady who sat beside me, whether she could see any of the curious things which he describes—the light emitted by crystals, for example? Here is the conversation which followed, as extracted from my notes, written on the day following the *séance*.

Medium.—'Oh, yes; but I see light around all bodies.'

I.—'Even in perfect darkness?'

Medium.—'Yes; I see luminous atmospheres round all people. The atmosphere which surrounds Mr. R. C. would fill this room with light.'

I.—'You are aware of the effects ascribed by Baron Reichenbach to magnets?'

Medium.—'Yes; but a magnet makes me terribly ill.'

I.—'Am I to understand that, if this room were perfectly dark, you could tell whether it contained a magnet, without being informed of the fact?'

Medium.—'I should know of its presence on entering the room.'

I.—'How?'

Medium.—'I should be rendered instantly ill.

I.—'How do you feel to-day?'

Medium.—'Particularly well; I have not been so well for months.'

I.—'Then, may I ask you whether there is, at the present moment, a magnet in my possession?'

The young lady looked at me, blushed, and stammered,

'No; I am not *en rapport* with you.'

I sat at her right hand, and a left-hand pocket, within six inches of her person, contained a magnet.

Our host here deprecated discussion, as it 'exhausted the medium.' The wonderful narratives were resumed; but I had narratives of my own quite as wonderful. These spirits, indeed, seemed clumsy creations, compared with those with which my own work had made me familiar. I therefore began to match the wonders related to me by other wonders. A lady present discoursed on spiritual atmospheres, which she could see as beautiful colours when she closed her eyes. I professed myself able to see similar colours, and, more than that, to be able to see the interior of my own eyes. The medium affirmed that she could see actual waves of light coming from the sun. I retorted that men of science could tell the exact number of waves emitted in a second, and also their exact length. The medium spoke of the performances of the spirits on musical instruments. I said that such performance was gross, in comparison with a kind of music which had been discovered some time previously by a scientific man. Standing at a distance of twenty feet from a jet of gas, he could command the flame to emit a melodious note; it would obey, and continue its song for hours. So loud was the music emitted by the gas-flame, that it might be heard by an assembly of a thousand people. These were acknowledged to be as great marvels as any of those of spiritdom. The spirits were then consulted, and I was pronounced to be a first-class medium.

During this conversation a low knocking was heard from time to time under the table. These, I was told, were the spirits' knocks. I was informed that one knock, in answer to a question, meant 'No;' that two knocks meant 'Not yet,' and that three knocks meant 'Yes.'

In answer to a question whether I was a medium, the response was three brisk and vigorous knocks. I noticed that the knocks issued from a particular locality, and therefore requested the spirits to be good enough to answer from another corner of the table. They did not comply; but I was assured that they would do it, and much more, by-and-by. The knocks continuing, I turned a wine-glass upside down, and placed my ear upon it, as upon a stethoscope. The spirits seemed disconcerted by the act; they lost their playfulness, and did not recover it for a considerable time.

Somewhat weary of the proceedings, I once threw myself back against my chair and gazed listlessly out of the window. While thus engaged, the table was rudely pushed. Attention was drawn to the wine, still oscillating in the glasses, and I was asked whether that was not convincing. I readily granted the fact of motion, and began to feel the delicacy of my position. There were several pairs of arms upon the table, and several pairs of legs under it; but how was I, without offence, to express the conviction which I really entertained? To ward off the difficulty, I again turned a wine-glass upside down and rested my ear upon it. The rim of the glass was not level, and my hair, on touching it, caused it to vibrate, and produce a peculiar buzzing sound. A perfectly candid and warm-hearted old gentleman at the opposite side of the table, whom I may call A., drew attention to the sound, and expressed his entire belief that it was spiritual. I, however, informed him that it was the moving hair acting on the glass. The explanation was not well received; and X., in a tone of severe pleasantry, demanded whether it was the hair that had moved the table. The promptness of my negative probably satisfied him that my notion was a very different one.

The superhuman power of the spirits was next dwelt upon. The strength of man, it was stated, was unavailing in opposition to theirs. No human power could prevent the table from moving when they pulled it. During the evening this pulling of the table occurred, or rather was attempted, three times. Twice the table moved when my attention was withdrawn from it; on a third occasion, I tried whether the act could be provoked by an assumed air of inattention. Grasping the table firmly between my knees, I threw myself back in the chair, and waited, with eyes fixed on vacancy, for the pull. It came. For some seconds it was pull spirit, hold muscle; the muscle, however, prevailed, and the table remained at rest. Up to the present moment, this interesting fact is known only to the particular spirit in question and myself.

A species of mental scene-painting, with which my own pursuits had long rendered me familiar, was employed to figure the changes and distribution of spiritual power. The spirits, it was alleged, were provided with atmospheres, which combined with and interpenetrated each other, and considerable ingenuity was shown in demonstrating the necessity of *time* in effecting the adjustment of the atmospheres. A re-arrangement of our positions was proposed and carried out; and soon afterwards my attention was drawn to a scarcely sensible vibration on the part of the table. Several persons were leaning on the table at the time, and I asked permission to touch the medium's hand. 'Oh! I know I tremble,' was her reply. Throwing one leg across the other, I accidentally nipped a muscle, and produced thereby an involuntary vibration of the free leg. This vibration, I knew, must be communicated to the floor, and thence to the chairs of all present. I therefore intentionally promoted it. My attention was promptly drawn to the motion ; and a gentleman beside me, whose value as a witness I was particularly desirous to test, expressed his belief that it was out of the compass of human power to produce so strange a tremor. ' I believe,' he added, earnestly, ' that it is entirely the spirits' work.' ' So do I,' added, with heat, the candid and warmhearted old gentleman A. ' Why, sir,' he continued, ' I feel them at this moment shaking my chair.' I stopped the motion of the leg. ' Now, sir,' A. exclaimed, ' they are gone.' I began again, and A. once more affirmed their presence. I could, however, notice that there were doubters present, who did not quite know what to think of the manifestations. I saw their perplexity ; and, as there was sufficient reason to believe that the disclosure of the secret would simply provoke anger, I kept it to myself.

Again a period of conversation intervened, during which the spirits became animated. The evening was confessedly a dull one, but matters appeared to brighten towards its close. The spirits were requested to spell the name by which I was known in the heavenly world. Our host commenced repeating the alphabet, and when he reached the letter ' P ' a knock was heard. He began again, and the spirits knocked at the letter ' O.' I was puzzled, but waited for the end. The next letter knocked down was ' E.' I laughed, and remarked that the spirits were going to make a poet of me. Admonished for my levity, I was informed that the frame of mind proper for the occasion ought to have been superinduced by a perusal of the Bible immediately before the *séance*. The spelling, however, went on, and sure enough I came out a poet. But matters did not end here. Our host continued his repetition of the alphabet, and the next letter of the name proved to be ' O.' Here was manifestly an unfinished word ;

and the spirits were apparently in their most communicative mood. The knocks came from under the table, but no person present evinced the slightest desire to look under it. I asked whether I might go underneath; the permission was granted; so I crept under the table. Some tittered; but the candid old A. exclaimed, ' He has a right to look into the very dregs of it, to convince himself.' Having pretty well assured myself that no sound could be produced under the table without its origin being revealed, I requested our host to continue his questions. He did so, but in vain. He adopted a tone of tender entreaty; but the 'dear spirits' had become dumb dogs, and refused to be entreated. I continued under that table for at least a quarter of an hour, after which, with a feeling of despair as regards the prospects of humanity never before experienced, I regained my chair. Once there, the spirits resumed their loquacity, and dubbed me 'Poet of Science.'

This, then, is the result of an attempt made by a scientific man to look into these spiritual phenomena. It is not encouraging; and for this reason. The present promoters of spiritual phenomena divide themselves into two classes, one of which needs no demonstration, while the other is beyond the reach of proof. The victims like to believe, and they do not like to be undeceived. Science is perfectly powerless in the presence of this frame of mind. It is, moreover, a state perfectly compatible with extreme intellectual subtlety and a capacity for devising hypotheses which only require the hardihood engendered by strong conviction, or by callous mendacity, to render them impregnable. The logical feebleness of science is not sufficiently borne in mind. It keeps down the weed of superstition, not by logic but by slowly rendering the mental soil unfit for its cultivation. When science appeals to uniform experience, the spiritualist will retort, ' How do you know that a uniform experience will continue uniform ?   You tell me that the sun has risen for six thousand years: that is no proof that it will rise to-morrow; within the next twelve hours it may be puffed out by the Almighty.' Taking this ground, a man may maintain the story of ' Jack and the Beanstalk ' in the face of all the science in the world.   You urge, in vain, that science has given us all the knowledge of the universe which we now possess, while spiritualism has added nothing to that knowledge. The drugged soul is beyond the reach of reason. It is in vain that im-postors are exposed, and the special demon cast out. He has but slightly to change his shape, return to his house, and find it ' empty, swept, and garnished.'

Since the time when the foregoing remarks were written I have been more than once among the spirits, at their own invitation. They do not improve on ac-quaintance. Surely no baser delusion ever obtained dominance over the weak mind of man.

END OF THE FIRST VOLUME.